WORKING WITH FARMERS
FOR BETTER LAND HUSBANDRY

Working With Farmers For Better Land Husbandry

Edited by NORMAN HUDSON and RODNEY CHEATLE

with
ADRIAN WOOD and FRANCIS GICHUKI

Intermediate Technology Publications
in association with
World Association of Soil and Water Conservation
1993

Intermediate Technology Publications Ltd
103–105 Southampton Row, London WC1B 4HH, UK

© Intermediate Technology Publications 1993

British Library Cataloguing in Publication Data
Working with Farmers for Better Land Husbandry
  I. Hudson, N.W.
  338.1

  ISBN 1–85339–122–0 (paperback)
  ISBN 1–85339–160–3 (hardback)

Typeset by J&L Composition Ltd, Filey, North Yorkshire, UK
Printed by SRP, Exeter, UK

Contents

SETTING THE SCENE

Preface N.W. HUDSON	3
Making Conservation Farmer-Friendly M.G. DOUGLAS	4
Soil and Water Management for the Nineties – New Pressures, New Objectives N.W. HUDSON	15

CURRENT RURAL DEVELOPMENT IN AFRICA

Introduction F.N. GICHUKI	23
Soil Conservation: an Ethiopian experience B.W. AREGAY and P.A. CHADHOKAR	23
Sustainable Soil and Water Management in Malawi L.A.C. BWEYA and N.J. MULENGA	26
Soil and Water Conservation in Tanzania F.B.S. KAIHURA and J.G. MOWO	29
Soil and Water Management in Uganda E.S. TAMALE	34
Soil and Water Management in Zambia N. MUKANDA and R. MWIINGA	37
Soil and Water Conservation in Zimbabwe S. KAVALO and G. NEHANDA	41
Current Aid Agency Approaches to Soil and Water Conservation P. EWELL, D. HUGHES, D.W. SANDERS and R. GALLAGHER, J.K. RANSOM, A. WOOD, C.S. WORTMANN	45
Donor Perspectives JOHN LYNAM	54

SOCIO-CULTURAL ISSUES

Introduction EVA TOBISSON	59
Changing Roles for Rural Sociologists EVA TOBISSON	59

Integrating a Socio-economic Perspective into Soil and
Water Management in Zambia 64
A.J. SUTHERLAND

ECONOMIC ISSUES

Economic Considerations for Participatory Development of
Natural Resources 71
J.P. HUNTER

Economic Management of Natural Resources by Rural
Communities 77
N. REYNOLDS

PARTICIPATORY APPRAISAL, PLANNING, AND DEVELOPMENT

Introduction 85
R.J. CHEATLE

Participatory Rural Appraisal 87
ROBERT CHAMBERS

Sustainable Small Farm Development – Frontiers in Participation 96
ROBERT CHAMBERS

Participatory Rural Appraisal for Agroforestry 101
M. AVILA

A Multi-disciplinary Approach to Socio-economic
Constraints and Research Priorities 104
A.J. SUTHERLAND and L.P. SINGOGO

OPPORTUNITIES AND BENEFITS

Introduction 113
N.W. HUDSON

Choosing Conservation Measures for Cropland on
Smallholdings in Kenya 117
D.B. THOMAS

A Strategy for Better Land Husbandry at Thabana Morena 126
T.F. SHAXSON

Smallholder Adoption of Some Land Husbandry Practices
in Kenya 130
R.J. CHEATLE and S.N.J. NJOROGE

Cash Incomes and Conservation: increasing both
simultaneously 141
S. CAIGER

COMMUNITY DEVELOPMENT CASE STUDIES

Community Participation 153
P. VEIT

Erosion Control in Machakos, Kenya 155
B. THOMAS-SLAYTER, C. KABUTHA, and R. FORD

Conservation in Bariadi, Tanzania 159
THE REV. H. GAPPA

Community Soil Conservation in Kabale, Uganda 162
E.M. TUKAHIRWA and P. VEIT

Agroforestry by Mobisquads in Ghana 165
C. DORM-ADZOBU, O. AMPUDU-AGYEI, and P. VEIT

Water Harvesting in Darfur, Sudan 169
YAGOUB A. MOHAMED

Joint Energy and Environment Projects (JEEP): wood energy conservation by participatory activities 172
RUTH KIWANUKA

Traditional Water Management and Irrigation Systems in Tanzania 174
O. MASCARENHAS

Rehabilitation and Construction of Earth Dams in Swaziland 177
F. MDLULI

LAND HUSBANDRY CASE STUDIES

Introduction 183
N.W. HUDSON

A Programme for Farm Improvement with Soil Conservation in Lesotho (FISC) 183
GEDION SHONE

Mobilizing Farmers to Counter Desertification in Nyanza District, Kenya 190
SR DOLORES RAUCH

Farmer Adoption of Improved Water Management on Vertisols in Semi-arid South-east Zimbabwe 197
P. NYAMUDEZA, E. MAZHANGARA, T. BUSANGAVANYE, and E. JONES

An Agroforestry Extension Project in Mazabuka District, Zambia 203
YEMBO KAONGA and ELIZABETH MALAYISHA

A Study of the Effects of Land Use on Water and Soil Resources on the Slopes of Mount Kenya 208
HANSPETER LINIGER

Developing Sustainable Grazing Associations in Lesotho   212
J.P. HUNTER and L.C. WEAVER

CONCLUSIONS

Next Steps Towards Better Land Husbandry   223
R.J. CHEATLE

Improved Training Approaches for Sustainable Land Husbandry   231
K.H.M. SEGERROS and R.J. CHEATLE

APPENDIX

The Workshop Background   241

Opening Address in Tanzania   244
DR B. MOSHI

Opening Address in Kenya   245
MR J.T. ARAP LETING

Closing Speech in Kenya   247
MR C.R.J. NYAGA

The Role and Contribution of Rural Sociologists to Sustainable Soil and Water Management   248

The Role of Economics in the Sustainability of Smallholder Systems   250
J.P. HUNTER and N. REYNOLDS

Participatory Approaches to Soil and Water Conservation for Sustainable Smallholder Development   253

References   256
Further Reading   261
Participants and Authors   262
Abbreviations Used in the Text   266
Index of Place Names   269
Subject Index   270

Setting the Scene

Preface

N.W. HUDSON

Many people in Africa face enormous problems as they struggle to feed themselves, and to generate sufficient cash income to meet the most basic needs. Their difficulties escalate as increasing populations work, within a decreased area, upon land that degrades continuously. Drought, civil war and repressive governments compound the problems in many countries. At best, the overall results are static crop yields and widespread poverty. Famine, refugee problems and abandoned lands are common symptoms of our failure to practise good land husbandry, and to manage our affairs to assure a decent livelihood for the majority of rural people.

We must search out solutions. In *The Greening of Africa*, Paul Harrison wrote: 'It is not the purpose of this book to pretend that the battle is well on the way to being won. That is far from the case. By and large the breakthroughs are surrounded and vastly outnumbered by failures. On present trends, disaster will carry the day. But it need not be that way. Our success stories are like seeds. If they are sown widely enough, they can take over the fields.'

Continuing that message, this book now reports on more seeds of hope and lessons for the future – ideas and results of activities which were introduced and debated at a workshop held in Tanzania and Kenya in 1991, described in the appendix to this book.

Some issues in agricultural development which were new and challenging five years ago are now accepted wisdom. For example:

- Land degradation is better controlled through improved land husbandry than by engineering-based soil conservation, which only tinkers with the symptoms
- Top-down programmes do not work well. The need is to work from the ground up, this means the involvement of farms and families and community groups at all stages, and
- Rural change is slow. Time, patience, and the matching of resources to real needs at the grassroots are all essential ingredients.

Now, there is a need to publish the accumulating experiences of ways to mobilize communities and stimulate self-help schemes. Each of the case studies reported here is unique but nevertheless, certain themes keep surfacing:

- The positive relationship between poverty, human and animal populations, and land degradation
- Better land use, with simultaneous increased production, can lead to higher incomes and better conservation
- The importance of organic recycling, vegetation cover and other aspects relating to the management of organic matter, to sustain soil fertility and stop degradation

- The importance of building on existing social structures, cultural patterns, and the traditional knowledge and practices of rural people
- The importance of women in agriculture, particularly the potential which can be released by eliminating the drudgery of carrying water and gathering fuelwood, and
- The effectiveness of NGO projects, especially as pioneers or catalysts, but the lessons will be most effective if projects support national policies and operate in co-operation with line departments.

This book first sets the scene with an outline of present approaches to soil and water management in different countries and aid agencies. Then follows a discussion of social, cultural and economic issues, and the mechanics of participatory working. The case studies reported are in three groups. There are examples of the opportunities and benefits of improved farming methods, case studies where an important part of the experience is stimulating community developments, and case studies with more emphasis on the methods and techniques used to promote better land husbandry. The last section discusses how to carry the wave of enthusiasm generated at the workshop forward. There are suggestions on vanguard research to develop methods for effective conservation farming, as well as ways to provide training and experience in the new ideas for fieldworkers. The appendix contains working papers from the workshops.

Making Conservation Farmer-Friendly

M.G. DOUGLAS

Land degradation: a continuing problem

A recent Food and Agricultural Organization publication (FAO, 1990) describes Africa's croplands, savannah, bush and forests as being under attack. They are seen as suffering from poor land management resulting in rapid land degradation. Although it is recognized that soil and water conservation practices can substantially contribute to reversing this degradation, the performance of past and ongoing soil and water conservation programmes and projects has, in most cases, been disappointing (IFAD, 1986, Hudson, 1991). The physical achievements of past efforts to tackle land degradation, in terms of areas treated and the range and number of earthworks constructed, are often impressive. However, such programmes have proved expensive to implement and rarely succeed in having any lasting impact on the problem.

The need to question past approaches and assumptions

There is a need to ask why, after 30 to 50 years of effort in Africa, is the area of land adversely affected by land degradation still increasing? Too often the blame is placed on the farmers who are accused of being ignorant, uncooperative, conservative, and unwilling to change. In other cases extension workers are accused of not doing their jobs properly, thereby failing to convince farmers of the benefits of soil conservation.

Recently, those involved in conservation programmes have started to recognize that the answer to the problem might be that their planning approach has been wrong, and the recommended practices unacceptable. There is a growing realization, and with it grudging acceptance, that farmers are behaving rationally when they reject the claimed benefits of conservation. This rejection is not surprising when projects are designed by outside so-called experts who fail to consult and involve farmers in the planning process. All too often the problems targeted (for example, soil erosion) are not perceived by the intended beneficiaries as being immediate priorities.

Likewise, it is now increasingly apparent to those working with small-scale farmers, that farmers are far from being conservative land users, nor are they reluctant to change their traditional farming practices when there are benefits from doing so (Sands, 1986). There is also a growing body of evidence to show that farmers constantly experiment, adapt, and innovate, with the aim of making adaptive improvements to their farming systems (Richards, 1985, Chambers *et al.*, 1989). If farmers fail to adopt conservation recommendations, it is not usually from ignorance, but because they think they are wrong. In some cases, farmers believe that a recommendation will not do the job it is intended to do. In other cases they are deemed inappropriate to their family's needs and farming circumstances (Shaxson *et al.*, 1989).

The need to redefine the problem

Soil conservation, *per se*, is not a priority goal for small-scale resource-poor farmers. Although they do not deliberately set out to degrade their land resources, they are often obliged to pursue land use practices, irrespective of whether or not they can be sustained, that best meet their family's immediate needs for food, fuel, shelter, and cash, as well as to meet their social and cultural obligations to the community in which they live (Douglas, 1989).

The implementation of successful soil conservation programmes is dependent on reconciling the different perspectives of the farmers and the land use and development planners as regards the severity and nature of the problems, and appropriate ways of overcoming them (Douglas, 1991b). Redefining the problem is a way for the planners to reconcile their differences with farmers.

Key concepts for successful conservation

Tackling soil conservation problems successfully requires acceptance of two key concepts when designing soil conservation programmes intended for small-scale farmers (Douglas, 1991b). First, that soil conservation practices should be 'farmer-friendly'. Second, that soil conservation planning requires a 'farmer first' approach.

To be successful, soil and water conservation has to be carried out voluntarily by farmers as part of their day-to-day farming activities. Practices need to be easily adopted and offer tangible benefits – that is they are farmer-friendly. This requires a bottom-up, farmer first, approach, involving farmers in identifying the problems and taking the decisions as to how they are to be overcome. This process ensures that conservation activities conform to the perceived needs and wishes of the farming community.

Limitations of runoff control structures

For many years conservation programmes have been based on the assumption that runoff is the main cause of erosion and that runoff and erosion are almost inevitable consequences of using land for crop production (Shaxson, 1989). The problem has been perceived as how to control runoff on agricultural lands to prevent loss of soil through accelerated erosion. Past conservation projects have, not surprisingly, relied heavily on the construction of conservation banks, storm drains, and artificial waterways to control run off and prevent soil movement. There is now an awareness that there are disadvantages to relying on structures alone because: (after Shaxson, 1989, Douglas, 1991a)

- Conservation structures have high direct costs for both initial construction and annual maintenance
- They may involve foregone costs by taking strips of land out of crop production, without necessarily producing any immediate benefit to compensate for the reduction in cropped area
- They can only counter the effects of runoff – they have no effect against raindrop impact
- They catch water and soil in bands along their uphill sides, but they do not prevent surface soil movement, nor promote rainwater infiltration, in the interbank areas where the crops are grown, and
- They can prevent the formation of gullies, but have no effect on declining soil fertility as a result of continuous cropping in the inter-bank areas.

Conservation banks are a way of dealing with the problem of excess runoff from unusually large storms. But on their own, they cannot substitute for improved conditions of soil structure, and cover in the interbank areas. Banks can only be used safely and effectively in support of better crop and livestock husbandry (Shaxson, 1989).

Focus on productivity loss rather than soil loss

Taking a farmer first approach requires the identification and definition of what farmers, not planners, perceive as the real problem. For most farmers the biggest problem is how to maintain and increase production (of crops, livestock, trees, and so on) on a sustainable basis for the benefit of their family, using the limited resources of land, labour, capital, equipment, and management skills, available to them. The implications of this for soil conservation programmes is that the focus should be on combating productivity losses, rather than preventing soil loss.

Soil erosion is a contributory factor in declining productivity. It should be seen as a consequence of how the land is used, not as the primary problem. Understanding why damaging erosion and runoff occur reveals they are foreseeable ecological consequences of inadequate, or inappropriate, land use (Shaxson, 1988). From this it becomes clear that efforts should be concentrated on understanding why farmers use the land the way they do, rather than determining the physical risks of erosion. Providing farmers with the means to make improvements in their land management practices is much more beneficial than concentrating efforts on promoting conservation measures that seek to combat erosion and runoff once they have begun. Likewise, identifying which of the farmers' existing practices are beneficial from a conservation point of view and which are not, and their reasons for pursuing them, will provide the basis for developing improved land management practices (Douglas, 1989).

Focus on improving land husbandry practices for soil conservation

It is no longer appropriate to preach to farmers that they must conserve their soil before they can get better crops. By redefining the problem the philosophy underlying soil conservation becomes that of promoting improved soil conditions for the benefit of root growth and plant production, and the conservation of soil and water. In the past, little emphasis has been put on improving husbandry as a way of achieving conservation, whereas, in reality, improvements in a farmer's crop, livestock, and land husbandry practices will be more effective in conserving soil and water than the implementation of physical conservation works alone (Shaxson, 1988).

The concept of husbandry is widely understood when applied to crops and animals. As a concept signifying understanding, management, and improvement, it is equally applicable to land (Shaxson *et al.*, 1989). Soil conservation has a somewhat negative image for farmers, due to its implied need to control and restrict present land use so as to preserve the soil for the future. The concept of land husbandry should prove more acceptable to farmers as it has a more positive image of managing and improving the present use of the land for productive purposes on a sustainable basis – that is, husbanding the soil resource.

Conserving soil by promoting improved crop and livestock management

Minimizing rainfall erosivity by improvements in cover and soil structure has greater effects on the overall processes of erosion and runoff than attempts to control runoff once it has occurred. In agricultural situations, the most versatile means of providing protective ground cover is to rely on the canopy provided by the leaves of well-grown crops, fodder species, and trees. Improvements in crop husbandry practices, such as early planting and at the optimum density, use of improved seed and fertiliser, and leaving crop residues on the surface, can significantly reduce splash erosion and improve water infiltration (Douglas, 1988 and 1991b). Likewise, the integration of livestock into the farming system by planting pasture as part of a crop rotation, using crop residues as feed and livestock bedding and applying the manure to the land, is a practical and profitable way for farmers to improve the surface condition of their soil and maintain organic levels (Douglas, 1988).

So, improved crop management (increasing ground cover), improved soil management (increasing organic matter levels and erosion resistance), and improved rainwater management (reducing splash and increasing infiltration) not only become more significant for the soil conservation specialist than soil loss management by mechanical control of runoff and sediment traps, but from the farmers' point of view they have direct production benefits in higher yields (Hudson, 1988).

The objectives and activities of conservation projects should no longer be targeted exclusively on traditional soil and water conservation concerns (for example, combating erosion, reducing downstream siltation). They should focus on methods of maintaining and enhancing the ability of the soil to sustain production. Soil conservation needs to be promoted within the context of agricultural development activities. It must not be undertaken as a separate programme. Soil conservation specialists need to be integrated as subject matter specialists in the government line departments responsible for agricultural development and extension activities. In this way the conservation dimension can become an integral part of the agricultural extension message disseminated to farmers (Douglas, 1991b).

Need for a farming systems development approach

Redefining the problem enables conservation objectives to be achieved at the same time as farmers' production objectives are satisfied. This can be done through the adoption of a farming systems development approach, instead of the traditional land management/erosion control approach. The aim is to integrate conservation into the existing farming system, rather than producing detailed conservation farm plans. The focus should be on the needs, constraints, and circumstances of the household members, not just the agro-ecological characteristics of the land holding, the physical risks of erosion, or the environmental requirements of the component land use enterprises.

Farming systems research (FSR) programmes have been undertaken for many years in eastern and southern Africa. This has been seen as a complement to station-based research. It is also a way of overcoming some of the weaknesses of on-station research in producing technical packages that are appropriate to the needs of small farmers operating in particular rural environments. Typically, the approach uses small interdisciplinary teams, usually composed of agronomists and economists, to obtain knowledge and understanding of farmers' circumstances within a specific area, to identify their problems and potential technological solutions, and to select a few of these for on-farm experimentation.

Although the approach, adopted by many of the agricultural research departments in east, central, and southern Africa, has much to offer the land use planner, much of the work conducted so far falls short of what is needed for promoting sustainable land use. Given that FSR has usually been conducted by small teams of agronomists and economists, the tendency has been to concentrate on the agronomic and financial aspects of crop production. It is clear that conservation concerns are only likely to be given due prominence when conservation specialists are included in farming systems teams.

Development of conservation farming systems

The development of conservation farming systems calls for an interdisciplinary team approach in which a range of disciplinary specialists (not just soil conservation specialists) work together. Each team member contributes from his or her individual technical perspective, with the team arriving at one common understanding of farmers' natural and socio-economic circumstances, and jointly developing appropriate conservation and land use recommendations (Douglas, 1989). The intention is to develop farming systems that are: (after IFAP, 1990)

- Productive in the short term, that is meet the farm family's immediate production goals of sufficient food to meet its consumption needs or a cash surplus or both, and
- Sustainable over the long term, that is do not destroy the natural resource base or deplete the capacity of the land to provide for the needs of present and future generations.

The outcome of a conservation farming systems development exercise may well be a set of research recommendations to test and validate different technical options. But the primary purpose is to identify, from a range of existing conservation-effective farming practices, those which can tackle the problems identified as farmers' priorities and are appropriate to their specific local circumstances.

In some cases, all that will be needed is to disseminate this information directly to the target farmers and local extension workers. This is the case where the recommended practices can be adopted within the existing resources of the farmers and extension support services. In other cases it may be necessary, before farmers can adopt particular conservation farming recommendations, to overcome local constraints and improve the necessary support services through project intervention (for example, provision of credit services, establishing input delivery systems and produce markets, tree nurseries, fodder banks, livestock distribution channels, farmer and staff training facilities). Here the farming systems development approach is an integral part of conservation project design (Douglas and Lai, 1989).

Farmers' land use practices are strongly influenced by the policy environment in which they operate. Therefore, an additional outcome from the analysis of the existing farming situation may be a set of policy recommendations. This is the case where there is a need to change the existing policy situation before particular technical recommendations can be adopted, for instance improving farmers' land tenure status, changing the producer prices to encourage or discourage the growing of particular crops, repealing legislation that regards all trees growing on farmers' land as the property of the government forestry service.

The need to recognize the constraints and circumstances under which small-scale farmers operate

Farmers, not government planners, ultimately decide what will, and will not, be done on their land holdings (Shaxson *et al.*, 1989). The development of conservation farming systems therefore requires a knowledge and understanding of the constraints and circumstances faced by farmers, as these will have influenced their past land use decisions and will affect future ones. It is important planners recognize that the constraints and circumstances faced by farmers will be area-specific, and the circumstances faced by small-scale farmers can be characterized in the following general terms (after Harrison, 1986, and Douglas, 1989).

Farmers are commonly faced with a range of adverse agro-ecological conditions.

Climatic Rainfall frequently occurs as short tropical storms with some forty per cent falling at erosive intensities. Rainfall distribution may be erratic with spells of drought an ever-present threat during the cropping season. And, in regions with distinct wet and dry seasons, a severe dry season will led to the death of annual vegetation and much bare soil exposed to the first rainstorms.

Soil Soils commonly have inherent fertility problems. Typically they are generally low in nutrients and organic matter, have high erodibility, and are often coarse textured with a low clay content and, therefore, low water-holding capacity. The more fertile soils are commonly associated with highland areas where steep slopes result in a high risk of erosion.

Farmers have access to limited arable land Pressure on land is usually high. This is caused by steadily increasing numbers of people and livestock. This leads to increasing use of marginal areas for food crop production, notably steeplands and semi-arid areas previously used for pastoral purposes.

Farmers are faced with a variety of socio-economic constraints Farmers have little in the way of formal education, and limited means of acquiring knowledge of improved farming practices. Their holdings are typically small in size and fragmented, without long-term security of tenure.

In addition, there is a shortage of cash for investment in the farm, and access to credit is usually limited. Farming is labour intensive, and yet labour is commonly in short supply, affecting the timeliness of crop husbandry operations. This is often further affected by inadequate draft power due to limited availability of draft animals and machinery.

Dependence of the household economy on agriculture requires that risk avoidance, rather than profit maximization, is the primary factor determining the farming practices used. Another constraint is that commercial farming is financially risky with producer prices commonly low, input costs high and market outlets unreliable.

Women farmers face additional constraints Women's dual duties and responsibilities to the family and farm lead to excessive burdens, for example seventy per cent of food production and one hundred per cent of the processing, child rearing, primary responsibility for wood and water gathering is performed by women. Although high and growing numbers of households are headed by women (thirty to sixty per cent), owing to male labour migration, men may retain the decision-making responsibility for land use/crop planning even when absent for most of the year. In addition women often have low civil status and power in the rural community, and therefore lack of access to extension advice and credit.

Limited government support to farmers Even when there is strong government commitment to agriculture, the extension, research, and conservation services are commonly short of funds and manpower. Also, the tax base is small, and subsidies for soil conservation paid during donor funded projects can rarely be sustained from government revenue budget sources.

On the basis of the above it is clear that conservation strategies based on high-input technologies involving complex engineering designs are inappropriate either financially or ecologically (often both) for small-scale farmers, that is they are not farmer-friendly.

Farmers should be offered a choice of practices

Studies have shown that small farmers rarely adopt complete technological packages. Instead, they tend to select from an array of introduced technologies and recommended practices those they perceive as most appropriate to the natural and socio-economic conditions in which they operate (Sands, 1986). The role of the planner in the farmer first approach should be to provide farmers with the basic principles – such as contour planting, use of hedge rows, rotations, ground cover – to offer them a range of options appropriate to the local area, and let them choose and experiment. This enables farmers to put together their own conservation farming package based on their specific needs. Farmers should also be given technical advice for the conservation layouts in their fields, for instance through teaching them to make and use A-frames or water levels for the alignment of contour hedgerows or grass buffer strips. In the past farmers often relied on a technician with a theodolite or dumpy level, or more recently the local conservation extension worker with a hand level.

The traditional approach to conservation has been to design a complete catchment layout of raised footpaths, diversion ditches, waterways, and the like, and to combine this with a set of land management recommendations (determined according to the local land capability classification) into a complete package (that is, conservation farm plan). Farmers' interests and needs vary, and what they actually want is a set of alternatives so they can choose for themselves according to their individual family requirements. To be farmer-friendly the conservation practices offered should ideally be:

- Simple – easy to demonstrate to farmers, and be understood and implemented by them
- Low-cost – within the financial reach of farmers, with limited labour requirements and no foregone benefits, such as land taken out of production
- Productive – lead to substantially increased benefits some fifty to one hundred per cent better than existing practices (that is, higher crop yields, increased fuelwood, guaranteed fodder supplies), preferably within the first year of adoption
- Sustainable – requiring limited effort or purchased inputs each year to maintain
- Low-risk – not susceptible to climatic variations or market fluctuations
- Flexible – leave scope for future developments. Farmers can change maize varieties after one season, but a decision to plant a particular species of long-lived perennial tree crop is not so easily reversed, and
- Conservation effective – contribute to the maintenance of soil productivity (for example, increase ground cover and soil organic matter levels, improve surface infiltration, reduce runoff, prevent surface movement).

The farmer first planning approach

The conventional approach to conservation has been to plan from the top down. Typically, the conservation expert travels out, identifies the problem in the field (usually perceived as loss of soil, gullying or downstream sedimentation), arrives at a solution with the aid of pre-determined technical guidelines, and only involves the farmers through an extension package at the implementation stage (Douglas, 1991b).

The farmer first approach aims to reverse this. Instead of starting with the knowledge, problems, analysis and priorities of the conservation specialists, it starts with the knowledge, problems, analysis and priorities of the farmers and their household members. In the Intermediate Technology publication *Farmer First* (Chambers et al., 1989) the main objective of the farmer first approach is described as: 'not to transfer known technology, but to empower farmers to learn, adapt, and do better;

analysis is not by outsiders – scientists, extensionists, or NGO workers – but by farmers assisted by outsiders; . . . what is transferred by outsiders to farmers is not precepts but principles, not messages but methods, not a package of practices to be adopted but a basket of choices from which to select'.

During the 1980s there was much innovative work related to people's participatory planning in a range of subject matter areas. There is a need to draw upon this for soil conservation purposes. Methods have been developed using rapid rural appraisal techniques to identify and analyse farmers' circumstances, diagnose their problems and design conservation-oriented solutions (Raintree 1987, Conway et al., 1987, Abel et al., 1989, Douglas, 1989). These offer starting points, but the analysis and identification of solutions is still primarily done by the experts.

Most of the reported examples of farmer first research activities (Chambers et al., 1989) relate to solving crop production problems, such as testing different cultivars, cropping patterns, and crop protection measures. The challenge is to find ways in which the conservation dimension can be incorporated into this approach, given that soil and water conservation as such is not a priority farmer goal and therefore of less immediate appeal. For instance, farmers may be willing to participate in screening programmes for the selection of preferred bean varieties by planting different varieties in small plots, but would be less interested in participating in a screening programme involving constructing different conservation works within their fields (Ashby et al., 1989).

Agroforestry practices, which have both production and conservation benefits, are likely to figure highly in the choice of options offered to farmers as the outcome of any conservation farming systems development programme. Experience from Kenya (Rocheleau et al., 1989) shows how the adoption of a farmer first approach led to the testing and adoption by farmers of both researcher-designed practices, and also innovative (and often more beneficial and practical) practices already used or suggested by farmers. An important feature of this programme was that participating farmers saw themselves as choosing, mixing and matching from a selection of possible agroforestry practices with some demonstrated feasibility, rather than adopting a proven package.

Promoting soil conservation as an integral part of a farming system, rather than a separate land management exercise, opens up the possibilities for farmer involvement in the development of improved land husbandry practices. Thus, farmers become involved indirectly in the development of conservation farming practices through their interest in sustaining and increasing production.

Community-level participation

The farmer first approach places primary emphasis on motivating individual farmers, and providing them with the means to adopt good land use and conservation practices within the boundaries of their own plots. But there are cases where it may prove possible, and necessary, to undertake conservation activities at the community level. This is where particular problems need to be tackled on a wider basis than the individual farm, or where there are tangible benefits to all members of a community which can only be achieved through a community effort.

Soil conservation projects have frequently called for people participation in implementing a package of conservation measures within the boundaries of a watershed or catchment area. Given that individual catchments may contain many farmers with separate holdings, and marked differences in farming skills, education, interests and needs, a conservation package that requires them all to work together, for the conservation of resources not solely their individual responsibility, is difficult to implement (Shaxson, 1981).

Participation is commonly seen as a voluntary contribution by people within the community in achieving the planners' conservation objectives (controlling runoff, reducing downstream sediment levels) and the project's predetermined targets for example, kilometres of conservation banks constructed, number of trees planted). Participation is typically a passive exercise being externally initiated and directed by project staff. At best, only the community leaders are involved in the problem analysis and project design stage. Their involvement is usually limited to providing information initially, and subsequently explaining the project to the community. This form of participation is essentially a short-term strategy for project implementation, with the community providing a temporary input of labour in order to produce the required physical outputs. The emphasis is on rapid mobilization of the community for their direct involvement in the task at hand, such as construction of banks, with the hope (rarely fulfilled) that the participatory spirit will continue when it comes to maintaining the structures after the project staff have withdrawn from the area.

Successful community participation in conservation projects requires that the community is involved from the start. Specifically in identifying the problems, determining the priorities for solution, selecting the practices to be adopted, and deciding how the project will be implemented. The need for participation may be the result of internal recognition within the community of the need to work together to tackle a problem, or the result of an external initiative. In either case, the role of the planners and project staff should be to assist, not direct, the farmers in the analysis, design, and implementation work.

Farmer first planning involving community participation requires recognition that there will be different perspectives on the nature, and extent, of the problems and possible solutions between the experts and the farmers. It must also be understood that individual groups within the farming community may perceive things differently, according to differences in their circumstances. When working with village communities it is important to ensure that the conservation activities agreed on are applicable to all the households, not just those making the decisions. Those in authority or positions of influence are usually male and better-off farmers. This necessitates that those doing the planning take the needs and circumstances of all groups who may be affected into consideration, for instance ensuring that the needs of women and landless households are also addressed (McCracken, 1990).

Conclusion

Making conservation farmer-friendly requires that soil conservation specialists should be familiar with classifying soils, analysing climate, and studying degradation processes, and know how to communicate and work with farmers in order to understand their circumstances and develop appropriate conservation farming practices. The immediate need is to develop new training programmes targeted at those responsible for soil conservation programmes covering such topics as how to: classify farming systems; identify farmers crop and livestock production problems; study the problems caused by degradation; analyse socio-economic data; and promote people's participatory research and development methods.

Specialists in extension and reseach services in subjects other than soil conservation will probably not have been directly involved in formulating conservation recommendations for small-scale farmers, or even feel that it is part of their responsibilities. Given the shift in emphasis from preventing soil loss to promoting improved soil productivity, it is clear that their disciplinary expertise is needed within the conservation systems development planning teams. Hence, the necessary training should be conducted with inter-disciplinary groups of subject matter specialists and not restricted to those with a technical background and responsibility for soil conservation. In this way it should be possible to promote a farmer first planning approach to soil conservation leading to the dissemination and adoption of farmer-friendly conservation practices.

Soil and Water Management for the Nineties – New Pressures, New Objectives

N.W. HUDSON

New factors and new problems

Human population
An increasing population means more people to be fed, clothed, and housed from a resource base which is shrinking. Any inbalance between demand and supply will inevitably worsen, as Malthus pointed out almost 200 years ago. It is inevitable that the population will continue to rise, because of the huge numbers of children in developing countries who will become reproductive adults in the next 10 years. Some continents are more

vulnerable to the problems of population pressure, especially Africa, where the current rate of population increase of 3.02 per cent corresponds to a doubling of the population in 23 years.

Livestock numbers
The data on livestock must be treated with caution. But, the evidence suggests there is little doubt that in many developing countries the livestock population is rising at least as fast as the human population. The levels of stocking recommended by range management specialists have never been accepted by the indigenous managers of livestock in most African countries. The stocking policy they usually follow is called opportunist stocking which means keeping high stocking rates to maximize production and growth in good years, at the expense of greater risk of losses in drought years. Whether this policy increases the risk of overgrazing and soil degradation is currently being debated, but whatever management strategy is adopted, the increase in livestock numbers is as inevitable as the increase in human population, which means more animals to be fed and more pressure on the land.

Political problems
There is currently much debate about whether the changes in climate are adding to the problem of food shortages in Africa. It can be argued that recent periods of drought are within the expected long-term variation. On the other hand, there is indisputable evidence from satellite imagery that the albedo is changing under grazing pressure.

Food supply and demand
The net result of these pressures is that world food production is increasing faster than population, so that production per person has increased over the last 40 years, but in Africa it has declined. There could not be a more pressing argument for better management of our land resources.

The place of research
The track record of national agricultural research in most developing countries is disappointing. The evidence for this criticism is that the usual strategy for conservation farming is a two-part programme – to build terraces, and to improve farming methods. What happens in practice, is that the terraces are built, usually assisted by subsidies or other incentives, but there is no tested package of practices to apply on the terraced land. The recommendations are likely to be limited to the obvious basic of use better seed, plant early, and use more fertilizer. The lack of applicable research results has in many African countries prompted donors to fund technical assistance projects to improve agricultural research.

It is worth looking at the reasons for the poor record. One is that developing countries have many demands on their limited budget, and agricultural research comes low on the list of priorities. Another difficulty is the institutional structure of research departments. Many developing countries have either inherited or copied colonial-type research institutions, where the pattern was developed to meet the needs of commercial farming, and the various farm operations are treated independently. So, on a typical research station there will be a department of livestock research, a department of pasture research, another on crops, and so on. The problem is that in subsistence farming all of these activities occur simultaneously, and on the same piece of ground. Thus the farming system cannot be improved by independent piecemeal advances in the components.

Single-component research, such as breeding, is best carried out on research stations where the variables can be controlled, but farming systems also need to be tested under real farm conditions. The current enthusiasm for farming systems research is in danger of overlooking the limitations of on-farm trials. These should be seen as extending and developing on-station research by additional factors such as farmer constraints or larger-scale operations. Currently, some writers seem to think that on-farm trials can entirely replace on-station research, overlooking the essential feature of research which is the ability to separate and control the variables.

All new technologies, and particularly farming systems, must be tested in real conditions to make sure they work, and to understand the risks of possible failure. In investment terminology, there must be no downside risk, and possible weaknesses must be eliminated before the system is recommended to the farmer.

In some African countries the increasing use of marginal land has led to new priorities in agricultural research so that more attention can be given to the special problems of land with low potential. The available technologies for soil and water conservation in semi-arid areas are reviewed in a recent publication from FAO (1987). Farming in semi-arid conditions inherently carries more risk of failure because the rainfall is more unreliable. The problems of using land also deserve more attention in research programmes because steeper slopes mean a greater risk of erosion, and require new technologies and new approaches which have not been adequately investigated.

New objectives and new ideas

National objectives, the policy

It is strange, but true, that few countries have a coherent policy for the use and development of their natural resources. But when a developing country does not have a long-term policy on land use, it is difficult to move on to the next stage, which is a strategic plan for the development of its land resource, and even more difficult to arrive at a tactical plan for implementation by the appropriate departments, divisions, or districts.

The starting point for any natural resource policy has to be an inventory of the resources. Nobody would start planning a steel industry without studies of the amount of iron ore and coal available. Similarly, it is no good trying to make a policy for agricultural development without knowing what resources are available.

Nowadays, it is easy to make an inventory of natural resources because of recent improvements in the techniques for acquiring data from the interpretation of air photographs and satellite imagery, and from field surveys. Today, only a small amount of fieldwork is required, compared with the time when parties of surveyors spent months in the field collecting data on soils, geology, and land use. After assembling an inventory of natural resources, the next step is to set a national **policy**, that is defining the long-term objective, stating the goal to be achieved in a defined period, perhaps 10 or 20 years. Policies are determined by national considerations, so they must be set at national level.

Next comes the **strategy**, which is deciding the general route to be taken towards the policy goal. The conditions influencing the selection of strategy will vary within the country, according to different physical and ecological conditions, so the strategy will be more specific, and perhaps apply only to a province or district. The timescale will be short, perhaps a five-year rolling plan.

The **tactics** are the details of how the strategy is to be implemented. This needs more flexibility so the tactical plan is usually over a short timespan, perhaps two or three years.

Programme objectives, the strategy
We can learn from the past and define what is required of a contemporary programme for better use and care of the land.

First, targets must be achievable. Politicians tend to make grand sweeping statements of intent which are impossible to achieve. Better to aim for a ten per cent increase which can be achieved, than to set the objective at one hundred per cent, and lose credibility when it does not happen.

Second, at last planners understand that the most important factor in developing programmes is the involvement and participation of the people. It has taken a long time to realize that trying to tell farmers what we think they should be doing has little effect. Effective programmes involve the people right at the start when objectives are defined.

Third, a successful programme cannot be built on a single component, such as introducing a new crop. All the other relevant factors have to be considered and included – the supply of inputs, availability of labour, market outlets and prices, and all the components of a farming system.

Fourth, change in farming systems is a slow process. Projects and programmes must stop looking for a quick fix in a few years. Most agencies now accept that 10-year programmes are more effective than the old three-year projects.

Fifth, programmes should aim to minimize external interventions, and to maximize the internal do-it-yourself activities.

Farmers' objectives
We now understand that the objectives of the commercial farmer to maximize yield and income are not necessarily those of the subsistence farmer. The latter is likely to be more interested in improving food security by decreasing the risk of failure, or improving the return on inputs of seed,

fertilizer, or labour, or reducing the labout input, or improving the quality of life by reducing drudgery, which is particularly relevant for the women farmers. We can define what new farming practices must offer in order to be enthusiastically adopted by farmers. First, there must be clear on-site benefits offering short-term improvements with large increments. Governments may appreciate the value of long-term benefits or down-stream benefits, or a ten per cent increase in average national yield. But none of these are important to the subsistence farmer although it is he or she who takes the decision on whether or not to implement the new recommendations.

We can further list what a new technology should and should not do:

- The technology must offer a quick pay-off. For crops, we should be thinking in terms of one year or less
- The technology must offer a high financial rate of return. A possible increase of ten to twenty per cent will not stimulate rapid up-take. This requires a return of fifty to one hundred per cent
- A new technology must not increase, and preferably reduce, the existing risk
- The technology must not require foregone benefits, particularly food. Practices which take land out of production, such as ridges or terraces, will only be acceptable if there are balancing benefits
- The technology must only demand low inputs of cash from the farmer
- A technology may be sound in technical terms but may still fail because it does not take account of special issues, for example the different roles of men and women in the farming system, or the attitudes towards communal systems, and
- A technology which is the development of an existing practice will be accepted more readily than one which is completely new.

Tactics, the technologies

Productivity In future, the incentive to farmers for improved soil management should be increased production. We should point out that better structure and more organic matter in the soil are aids to better production, instead of recommending them because they reduce soil erosion. Similarly, if development workers want to encourage better ground cover or the use of contour farming, or ridging, these practices will make more sense to the farmer if they are recommended as aids to better production, rather to reducing runoff. The use of more fertilizer, better seed, and rotations can also be related to improved yield instead of to soil conservation.

Extension Advice will probably be in the form of packages of practices which make up a farming system, rather than concentrated on cropping or grazing or woodland in isolation. Treating these as separate subjects is the product of the western scientific approach, which dictates that these practices are traditionally carried out on separate areas of land. But in peasant farming they are likely to exist on the same land. The difference between crop land and grazing land may be a question of time, rather than space, for example, where land is used for grazing after the crop has been harvested, and the tenure changes from individual rights to the cropping to communal rights to the grazing.

Water conservation   The new strategy will probably place more emphasis on water management and water conservation, and less on soil conservation. It is likely that the preferred approach will be to manage land for maximum infiltration and minimum surface runoff, in order to achieve better yields where soil moisture is a constraint. The fact that this simplifies the disposal of surplus runoff is a bonus or by-product.

Building on tradition   Wherever possible we should try to build on, and improve, traditional practices rather than introduce foreign methods.

Simple measures   It used to be assumed that there is abundance of cheap labour in the agricultural sector in all developing countries. We know now that there are countries in Africa where shortage of labour is a production constraint. In any case, nobody wants to spend more time than necessary labouring in the fields. Thus, there has been a move towards conservation measures which need less labour than terracing. Some are just simpler, such as grass strips and trash lines. Others only terrace part of the land, such as intermittent terraces or hillside ditches, while some measures spread the labour over several years, such as *fanya juu* terracing and other forms of progressive terracing.

Conclusions

Pressures on land resources will continue to increase. This will lead to increase risk of degradation, and more use of marginal land, particularly medium- and low-potential land and steep land. Therefore, applying the best soil and water management becomes increasingly important.

We have to do better than in the past. Particularly, we need carefully thought-out national plans for the best use and development of land, and strategic and tactical plans to implement this policy.

We need national agricultural research services to redirect their programmes towards present needs and problems, particularly multidisciplinary problem-oriented studies of small-scale farming systems.

The key to getting programmes in place on the ground is the involvement and willing participation of the land-holders. This means several changes in the emphasis of soil and water management:

- From projects trying to cure erosion to programmes which prevent it
- From thinking soil conservation to increased productivity and land husbandry
- From earth-moving soil conservation to better farming
- From quick-fix projects to long-term programmes, and
- From top-down plans to people-led development.

Current Rural Development
in Africa

Introduction

F.N. GICHUKI

This section reviews the state-of-the-art in soil and water conservation. The review is organized into three groups: country papers, agencies approach, and donors perspectives.

Country papers illustrate the challenges facing those involved in soil and water management problems, that is farmers, governments, non-government organizations, and donor agencies. Issues highlighted are:

- Institutional build-up and collaboration
- Training and extension
- Technology development and adoption
- Planning, implementation, and maintenance approaches
- Policy issues, and
- Conservation programmes.

The discussion of international agency approaches to soil and water management presents the experiences of FAO, IUCN, and three members of the CGIAR. Their roles complementing national soil and water conservation programmes, or national agricultural research programmes, and the main elements of their new programmes and past experiences are valuable to the development of soil and water management systems that are sustainable at the smallholder level.

The review of donor perspectives presents the role of donors, their interests and commitments, and priorities.

Soil Conservation: an Ethiopian experience

B.W. AREGAY and P.A. CHADHOKAR

The problem of soil erosion

The problem of soil erosion in Ethiopia is concentrated in the Ethiopian highlands, which comprise 536 000km^2 or forty-four per cent of the total area of the country. The main reason for accelerated soil erosion is over exploitation of natural resources. This has resulted from increasing demands of the growing human and animal populations. The exploitive land use practices include deforestation for expansion of cultivation, grazing, fuelwood, and timber. These reduce the protective plant cover, thereby exposing the soil to the forces of erosion. The history of human migration in Ethiopia from north to south can be related to the land degradation due to erosion and its effects on the productivity of the land. The frequent drought and famines are also partly related to land degradation.

Today, the highlands are the centre of most human activities and are fast deteriorating due to soil erosion. The highlands support around eighty-eight per cent of the human population and sixty-six per cent of the livestock population. Over ninety-five per cent of the land is intensively cropped and forms the basis of over ninety per cent of the country's economy (EHRS, 1986).

Recent estimates indicate annual soil loss in Ethiopia is between 1.5 and 3 billion tonnes (EHRS, 1986). Of this, about fifty per cent occurs in crop lands where soil loss has been reported to be very high (296 tonnes/ha/year on a sixteen per cent slope with a Tef crop (*Eragrostis abasinica*) on nitisols (SCRP, 1987). The EHRS study estimated that about fifty per cent of the highlands is already significantly eroded, of which about 14 million hectares are severely eroded. Two million hectares have reached a stage of irreversible destruction and cannot sustain cropping in future.

The forecast for the future is even gloomier. EHRS (1986) forecasts that seven per cent of the highlands will be bare rock, and eleven per cent will have less than 10cm depth, by the year 2010, compared to an estimated four per cent in 1985. If the present trends continue in the highlands as a whole, in addition to the present 20 000km^2 incapable of sustaining arable cropping, a further 76 000km^2 is likely to be similarly degraded. The loss in soil productivity is usually greater because erosion generally takes away most productive part of the soil. The effects of soil degradation have been described as:

- Productivity of land as well as production per unit area decreases
- Each family cannot meet its daily food requirements unless it cultivates more land to compensate for decreasing productivity
- Livestock concentration on the constantly decreasing area results in overgrazing, which reduces livestock productivity and plant cover
- The regulatory capacity of the mountains is drastically reduced, and
- The overall effect is frequent drought, famine, and related disasters.

Efforts to combat erosion

Ethiopia has a long history of following traditional conservation methods. There are numerous examples of certain parts of the country where these techniques can be seen. For example, stone terracing in Konso, Gamu Goffa, random bench terraces in north Shoa and Hararge, contour bench terraces and tied ridges in Konso, drainage furrows of north-east Shoa, and sod rotation, trash bunds, trash heap composting, and fallowing. To date, these techniques have not been evaluated, nor has there been any attempt to improve them or popularize them.

The scientific conservation programme is a recent phenomenon. A start was made in early 1970s, but serious attempts on a large scale were delayed until the early 1980s, when assistance of the WFP and UNDP/FAO became available. Initially, conservation work was concentrated in severely degraded areas with reclamation measures and forest planting. As the major thrust was on reclaiming degraded lands, scant attention was paid to preventive measures on productive lands. Thus the problem spread.

In spite of its recent beginning, and limited manpower and resources, a vast amount of conservation work has been done with WFP assistance in the relief areas. However, after 10 years of work, even this huge programme has only achieved seven per cent of the physical conservation measures required. Moreover, the concentration on terracing has not been accompanied by improved land use, and there is no voluntary adoption without WFP payments.

One of the most cost-effective, though slow, conservation activities has been area closure (Chadhokar, 1989). More recently, with the assistance from UNDP/FAO, there have been significant improvements in the manpower development, refinement of conservation techniques, introduction and refining of watershed planning and management, and most importantly, introduction of biological conservation measures.

Over the past few years, it has been realized that long-lasting soil and water conservation cannot be achieved only by following structural measures. It is now understood that farmers prefer simple, less costly techniques that can be easily integrated with the traditional farming practices. Forage grasses, legumes and legume fodder shrubs and trees have been used to stabilize soil bunds and gullies with limited success, due to free movement of livestock which graze these species and destroy them.

The government has started promoting a bottom-up approach to conservation. At the same time it is building institutions and training manpower at all levels to promote the development and widespread adoption of cost-effective conservation technologies.

Future consideration

The experience gained over the past few years has provided valuable information which can be applied to future programmes to make them more efficient and effective. The following considerations have emerged from past experiences:

- There should be a well-defined strategy on soil conservation, and the activities should be carried out after careful planning based on priorities and available resources
- More attention should be given to the causes of erosion, instead of the symptoms. Most of the causes can be eliminated by following conservation-based farming
- Afforestation should be linked with ownership of trees, as opposed to the current communal ownership of community forest resources
- Agropastoral and agronomic measures should be used in combination with structural measures to improve the effectiveness and improve land productivity
- Adequate attention should be given to water conservation in semi-arid areas, and
- Policies which improve the effectiveness of soil conservation programmes, such as training, extension, and legislation should be promoted.

Sustainable Soil and Water Management in Malawi

L.A.C. BWEYA and N.J. MULENGA

Introduction

Malawi is 118 576km^2 of which 94 276km^2 is highland. About thirty-three per cent of the country is arable land, and a further fifteen per cent could be successfully cultivated if intensive conservation measures were followed. The remaining land is not arable, mainly due to high relief, shallow soils, and unreliable rainfall. Malawi also has a considerable amount of water in the form of a large lake. Forest reserves, national parks and game reserves cover about twenty-one per cent of the total land area.

Increases in human population are leading to the over-exploitation of the land resources, and cultivation is extending into marginal areas. Currently, government policy on agriculture takes this increasing pressure on land into consideration, and is geared towards encouraging the people to practise intensive agriculture through good land husbandry.

A high percentage of the population lives in rural areas and depends on agriculture for its livelihood. Small farmers, particularly those operating close to the subsistence margin, can rarely afford most resources. Therefore, they are unwilling to give up the prospect of immediate production to ensure long-term benefits and maintain practices that enhance soil and water maintenance.

Available statistics show that cultivation had been expanding at approximately three per cent. Currently, there is very little arable land left.

Soil and water management problems

There are a number of problems related to soil and water conservation, ranging from erosion to pollution problems, and from technical to social problems. Population pressure on the land is causing problems for soil and water management – as the holdings get smaller and smaller (less than one hectare in some parts), so the resource-poor land user will over use it. In this scenario, some aspects of good management will be overlooked. Couple this with the fact that expansion of land for agriculture exploits areas which should perhaps have been left as catchment areas for water resources. The problem is aggravated by the practices that are being used for soil and water conservation. It is sometimes felt that the problems are due to a lack of appropriate extension messages. Many of the messages have taken the top-down approach with outside 'experts' deciding what the poor farmers should do to maintain or conserve the land without taking into consideration their own environmental knowledge, preferences, priorities and constraints. Often the practices and messages are alien to the poor farmer, so he or she does not adopt them. The methods used are often unsuitable for extension through groups. Thus, coverage of the farmers is limited, and messages are not reaching the target group at the right time. Lack of government resources means that the resource-poor farmer is left stranded, despite the fact that he or she wants to improve his practices.

Deforestation followed by cultivation on steep lands often cause land degradation. As a result the small farmer has to increase his or her cultivated land to produce enough food for the family.

Another cause of land degradation is overgrazing. Although this is a localized problem, it is worst in areas with large animal populations, such as the northern and southern tips of Malawi. Livestock numbers are too high to sustain on the existing pastures. However, because livestock are communally grazed, there has been no practical grazing system to alleviate this problem.

The influx of refugees from Mozambique into the border areas in central and southern Malawi has aggravated the problem of land degradation. There is a shortage of fuelwood and grazing as some of the refugees brought livestock with them. In addition, poor management practices encourage serious gully erosion. The farmers often ignite uncontrolled bush fires, and these influence high runoff in the river catchments.

National priorities

It is unfortunate there has been a large increase in the amount of land brought under cultivation, most of it unsuitable for arable use. Recommended methods of cultivation are often ignored in favour of farming practices that result in rapid deterioration in soil structure and fertility. Perhaps, this can only be explained in terms of the socio-economic circumstances of the resource-poor farmer. To develop a sustainable soil and water management programme we believe that:

- Land resources surveys are valuable for planning appropriate conservation practices for specific areas
- Land use planning and soil conservation should offer a range of technical solutions to farmers, so that they can select the most suitable for them
- Training should be aimed at backing-up small farmers' existing knowledge of soil and water maintenance
- Environmental education by lectures in schools and colleges merit consideration. Future smallholder farmers are at school-going age and they need to be prepared for the future, and
- Research in soil and water maintenance to identify practices which the smallholder farmers can adopt.

Soil and water management practices

The national policy on soil and water management is to ensure a balanced and sustainable utilization of land resources through rational management. This is to be achieved through provision of relevant training and skills in all aspects related to the management of soil and water, and promotion of research for soil and water management.

Integrating conservation into the farming system offers an opportunity to adopt soil and water management practices, without causing alarm or suspicion about the new technologies. Lip service is paid to people's participation, but it often just means getting the farmers to agree to do what the experts consider necessary. This is tragic and all too common.

Soil and water management practices vary from one agroclimatic zone to another. The common practices include agroforestry, strip cropping, grass and buffer strips, mixed or intercropping, and contour ridging. Soil and water management is usually through simple practices like burying crop residues during garden preparation by the smallholder farmers, and making compost from animal manure. The use of green manure through agroforestry practices is being promoted in Malawi (*Leucaena* hedgerows and *Acacia albida* leaves). It is also recommended that all crops should be grown on ridges. This is to control soil erosion and conserve water. Correct plant populations, established by research, play a large role, and this is promoted to the small farmers to aid soil and water conservation.

Research and extension linkage

The link between research and extension is important, because techniques developed by research end up with the farmer, and farmers' problems have to be sorted out by research. In Malawi, the link is clear. Adaptive research teams attempt to solve farmers' problems right in the farmers' gardens. If problems are more difficult, they are referred to the commodity research teams at a research station.

Research concerning tillage practices is given high priority. There is an urgent need to develop soil management measures that are not complicated so farmers can adopt them easily. This is important because a resource-poor farmer has to allocate his or her time to many activities. A recommended practice often fails because socio-economic circumstances were not taken into account.

Sustainable soil and water management is frequently discussed. But enabling a small farmer, who normally works fragmented land holdings, to participate is difficult because he or she is often preoccupied by family problems and meeting social and cultural obligations of the community. There is a need for regional data banks to supply information to farmers. Possible techniques could be tested on-farm, and the results fed back into the database. A regional information network is suggested, so that effort is not duplicated. It is believed that there is a lot of unexploited information within the region.

Project planning and sustainable soil and water management

Experience has shown that all projects require an environmental perspective. An environmental impact assessment must be done before approval can be given. Therefore, sustainable soil and water management are considered at the design stage of a project. Because of general concern, a Department of Research and Environmental Affairs was established in

1983. Strong inter-departmental linkages have been developed. The National Rural Development Programme in Malawi has encouraged multidisciplinary and interdisciplinary teamwork throughout the implementation of the programme. There have been both good and bad experiences. While discussions have been quite successful and general agreement reached at ground level, there have been problems because departments have faced staffing and funding difficulties. This has affected the smallholder, as most of the initiatives have been presented in a piecemeal fashion rather than as part of a unified programme. The experience shows that it is problematic to implement a programme on the ground when too many agencies are involved. Monitoring and evaluation becomes difficult, co-ordination ineffective, and assignments not fulfilled.

Conclusion

Agricultural production must start with the small resource-poor farmers, but to enable them to participate in sustainable soil and water management, practices must be simple and easily adoptable. There is much that can be learned to improve what is already being done. However, it is essential that the farmer does not regard the improved practice as alien to his or her farming system. Integrating conservation into the farming systems has a great deal of merit. One of the most important aspects is to fit improved technical practices to the circumstances of the farmer. Also, it is important to remember that small farmers are rational and purposeful in their behaviour, and change within their normal lives causes anxiety.

Whatever our intentions, whatever practices are available for sustainable management of soil and water, and whatever projects are identified, as long as we do not understand these small-scale farmers we shall be working in isolation. In Malawi, we try, as much as possible, to consider the socio-economic circumstances for effective sustainable soil practices. As yet we have not completely achieved the intended objective.

Soil and Water Conservation in Tanzania

F.B.S. KAIHURA and J.G. MOWO

Introduction

Tanzania is a country with an area of about 942 000km^2. It has a population of more than 23 million people growing at the rate of 3.3 per cent a year. About eighty-nine per cent of the population live in rural areas. Agriculture is the mainstay of the economy. It provides eighty per cent of exports and employs ninety per cent of the population. Except for the coastal belt, most of the country is located on the central African Plateau at 1000–1500m.

The pattern of land use in Tanzania varies. Highland areas with annual rainfall ranging from 1300 to 2000mm are usually good for agriculture. These areas are highly populated with over 150 people for each km^2. Highlands are renowned for cultivation of export cash crops like coffee, tea, pyrethrum, and food crops. Lowlands are generally drier with lower annual rainfall of 250 to 750mm, and higher temperatures. Traditionally, lowlands have been marginal areas supporting pastoralists and wildlife. However, population pressure in the highlands has begun to spill over into such areas.

The country has 88.6 million hectares of potentially cultivated land under rainfed conditions, but only six million hectares (sixteen per cent) are cultivated. A total of 933 000ha have potential for irrigation, but only 144 000ha are presently under irrigation, of which 72 000ha are under smallholder production.

Tanzania has about 44.2 million hectares of forest resources, 43.2 million of natural miombo woodland, 950 000ha of closed forest, 29 000ha of planted trees in woodlot and schools, 60 000ha of soft woods and 6000ha of hardwood in industrial plantations. The national demand for wood is 34 million m^3, and greatly outstrips the supply (19 million m^3) indicating an exploitation of natural forests to fill the gap.

Environmental problems

Environmental problems in Tanzania and elsewhere are the result of man's mismanagement. Agriculture and deforestation are the major factors contributing to the processes of environmental degradation.

Soil degradation
Population pressure, leading to increasing land shortage in the highlands, results in intensification of agriculture, loss of topsoil, and hence productivity is reduced. In the medium and lower potential areas, local population growth is augmented by increasing immigration from crowded high potential areas. Farming in marginal areas adds to the problem. Farmers neither adhere to conservation tillage nor relate climate, topography, and soil type to tillage practices, particularly on state-owned land. Runoff and sediment transport take away the top soil and make it less productive. Nutrient status is generally low for most soils in the area, but does not affect productivity at the smallholder production level.

Water resources degradation
Soil moisture stress is a major problem affecting productivity. This is typical in bimodal transitional rainfall areas, which experience a dry spell (10 to 20 days) during the long rains (March to June) which coincides with the reproductive stage of the crops, negatively affecting crop yields despite the use of fertilizers, chemicals, and other improved management practices. The short rains (November to December) are variable and erratic. They cause more damage than good, carrying away most of the topsoil in the form of runoff and sediment. The dry season is also longer (six to eight

months) than the rainy season. The crops suffer both moisture and nutrient stress.

Vegetation degradation
Stocking rates are often higher than the carrying capacity of the range lands, resulting in overgrazing. Consequently, gradual land degradation develops. This leads to the migration of people and stock to new areas, where the same process occurs again. Examples include stock movement from north eastern Tanzania (Mwanza and Shinyanga) to Mbeya in the south and from Dodoma region to Arusha and Morogoro regions.

Deforestation
The national demand for wood exceeds supply and results in deforestation. Increasing shortage of land for crop production also contributes to depletion of forests. In some areas in Tanzania, for example Sukumaland, cow dung and crop residues are used as sources of domestic fuel. Rangeland fires contribute to woodland destruction, which enhances erosion, biological changes in the root zone, and water deficits.

Government policy

Before 1980, Tanzania did not have an explicit policy concerning environmental management. In 1967 the Arusha declaration outlined how natural resources should be managed. The first document on soil conservation appeared in the second five-year development plan (1969 to 1974). In 1983, the National Agricultural Policy was established. It emphasized that soil conservation measures should be incorporated in all land use plans and village masterplans. In 1983 to 1985 a structural adjustment programme (SAP) provided the framework for the government's response to the growing economic crisis in the early 1980s. Agriculture and forestry were the major sectors addressed, with emphasis on smallholder producers. It was believed that maximum return on the use of scarce resources would come from a programme which focuses on smallholders, who account for eighty per cent of the agricultural output and forty per cent of GDP. The programme emphasized land use planning and soil conservation in the villages through:

- Provision of adequate land to the villages
- Incorporation of soil conservation in farming systems
- Demarcation of the country into land use zones
- Encouraging and assisting people to move from over populated areas to low density areas
- Introduction of modern animal husbandry in order to improve livestock production and remove overgrazing problems
- Establishment of land rehabilitation and soil conservation projects, and
- Village and urban afforestation.

The government established the National Environmental Management Council (NEMC) in 1983 to monitor and ensure proper utilization of environmental resources. In 1984 the Land Use Planning Commission (LUPC) was established to implement the 1983 agricultural policy in soil and water conservation. Both bodies are under the Ministry of Natural Resources and Tourism.

Soil and water conservation activities in Tanzania

Several ministries and agencies are involved in soil and water conservation, but their programmes are not co-ordinated, resulting in duplicating of work, and, at times, concentration of projects in one area. Government funds allocated to soil conservation projects are usually insufficient, and may not reach destined users in the field. Since allocated funds are often inadequate, conservationists do not get their share and remain idle in the field. Second, nowadays local governments emphasize conservation projects in the districts, but this emphasis is not reflected in the local government's budget. As a result, many projects without donor agencies have remained as paper documents.

Major obstacles for effective implementation of conservation projects include:

- Lack of cross-sectoral planning approach
- Lack of professional awareness of environmental issues
- Lack of trained manpower in environmentally related fields, and especially for identification, design and implementation of conservation projects
- No emphasis on socio-economic factors
- Lack of an integrated approach by multi-disciplinary teams, and
- Inefficient extension services due to limited personnel, transport and poor communication between research and extension.

Strategies for research and development

Farmers' participation

Excluding farmers' participation was a crucial setback in Tanzania following the 1945 to 1955 land use schemes. In most schemes, the colonial government ignored the fact that local people had more knowledge about their own environment than officials. For example, they embarked on the construction of large bench terraces on steep land in Mgeta in the Morogoro region, which the local population knew were ineffective, because they resulted in land slides after heavy rains. From a native point of view, overstocking is a necessary evil, essential to maintain a high standard of health and wealth. Increased stock is related to increased peace and security. De-stocking policies should therefore take these factors into consideration. Past, and even present, policies have failed to respect farmers' points of view, resulting in the failure of most projects.

Socio-economic aspects

Despite many research and development projects to improve the capacity of smallholders their development has stagnated. Farm products are not assured of good prices. In Tanzania, fixed prices on some crops are much lower than the actual market prices. Some products do not even reach the market. Farm inputs are expensive and unaffordable. The rural infrastructure is under-developed, and transport costs are very high. Labour is another constraint, particularly when enterprises are diversified to reduce risk.

Women in the household are the main source of farm labour. Men are only occasionally available as they seek off-farm employment to earn extra income to support the family. Hired labour is expensive as freelancing youngsters flee to towns for self-employed jobs. The smallholder's perception is that development projects will cost him or her extra labour, time and money. Planners have often overlooked this, resulting in unsuccessful projects. It is important to study the socio-economic implications in soil and water management and conservation projects, which can only be adopted after understanding the people and their culture, and by involving them in identification, design, and implementation of projects. By doing this projects will be set within the framework of their needs and existing technical and financial capabilities.

Farming systems research (FSR)
Farming systems research is an important approach to operational research in agriculture. It is often projected as an easier and cost-effective means of generating technology relevant to the means and requirements of the farmer. An integrated use of farm surveys, diagnostic experiments, and adaptive research on farmers fields by an inter-disciplinary team of social and natural scientists constitutes an important feature of the approach. Without substituting for, or competing with, other applied or basic research, FSR helps in picking-up elements of different disciplines and integrating them into relevant technology packages. Because of its bottom-up approach to technology generation, FSR begins with the farmer and ends with the farmer.

In Tanzania sociologists are scarce, and FSR should be considered as an alternative to promote technology adoption for soil and water conservation. In Tanzania the approach focuses mainly on crop varieties, and is currently being tested in soil and water management and conservation.

Co-ordination of soil and water management research

Different ministries are responsible for soil and water management and conservation in the country. Donor agents also carry out conservation projects under different ministries. In some cases, project areas are selected to meet the donor's interests. Many local and foreign projects, are often found in one area, and sometimes address the same problem. To avoid misuse of scarce resources, manpower and duplication of work, a cross-sectoral committee as a national co-ordination unit must be established. Currently, co-ordination is within individual ministries.

Soil and Water Management in Uganda

E.S. TAMALE

Introduction

Uganda is a small country situated along the equator. It has an area of approximately 243 000km^2, of which 50 000km^2 is Lake Victoria. Relief is complex, characterized by undulating plateaux with flat-topped hills, and a highland complex in the west. The annual rainfall averages 1000 to 1850mm. The major soil types are fertile. The depth of soils has been reduced in many parts of the country because poor farming practices induce soil erosion. A proportion of the country was covered by equatorial forest, but these are either gone or degraded, although some islands of forest remain. Savannah grassland dominates most parts of Uganda, and desert steppe vegetation is found in the arid areas of north-east Uganda.

The agricultural sector is dominated by peasant small-scale farmers who account for over ninety-five per cent of the total agricultural production. Agriculture contributes over ninety per cent of the gross domestic product, and provides employment to over ninety-two per cent of the population. Uganda is basically an agricultural country and the socio-economic advancement relates directly to the efficient use and management of the natural resources.

The problems

Uganda, once the Pearl of Africa, has over the past few decades experienced a rapid decline through degradation of the natural resource base. This has been greatly influenced by lack of proper planning, high population pressures, poverty, and lack of understanding. The increasing population and land scarcity have put greater demands on the resources, provoking people to cultivate areas of marginal productivity, such as steep slope natural waterways, swamps, and other types of wetlands.

The population pressures
The population of Uganda is about 18 million in the recently completed census, and the growth rate is estimated at over 2.9 per cent a year. Over ninety per cent of the population is rural, with an average density of 50 people for each km^2. The heaviest density is around the Lake Victoria region and in the highland areas of south-west Uganda.

Degradation of water resources
Over fifteen per cent of Uganda's total surface area is covered by open water and swamps, more or less evenly distributed in the country. Despite the abundance of water in the country, many people in rural areas lack access to safe water for domestic use. Women and children have to walk several kilometres to reach water. Unsafe sources, and water-borne diseases are common.

Uganda's water resources used for agriculture, industrial purposes, transportation, electric power generation, and recreation have been subjected to numerous mismanagement practices, such as pollution, swamp destruction, excessive fishing, and promotion of poor quality spring water. These practices have caused a number of environmental problems like water-borne diseases, siltation, and reduced biodiversity.

Soil degradation
The deterioration of Ugandan soils has been aggravated by misuse of land by smallholder farmers. Poor farming techniques and malpractices, for example indiscriminate elimination of trees from the farming systems, deforestation, overgrazing, destruction of watersheds, excessive burning of vegetation, and excessive tillage have all resulted in decline of soil fertility, loss of organic matter, soil acidification, low soil-water holding capacity, and low biological activity in soils. Associated with these changes are decline in crop yields, poverty, and low standards of living.

The consequences
There is an imbalance between human demands on the natural resources for survival, and the capabilities of eco-systems to meet the demands placed on them. This situation is compounded by a lack of proper planning, and of effective legislation for wetland and watershed protection. The situation is worsened by human abuses because of leaving the management and utilization of water resources in public hands and by the collapse in extension services, which means that most farmers in the countryside are not being advised about soil utilization and management techniques.

Strategies for soil and water management in Uganda
Short-term, and long-term strategies and plans illustrate the government's commitment to consider environmental protection as a critical ingredient of the national development.

Institution building
In 1986, the Ugandan government established the Ministry of Environment Protection as the first state organization responsible for environment protection. The ministry has embarked on a number of strategies aimed at resolving environmental problems that arise from the utilization and management of soil, water, and forest resources in the country. The ministry has three departments, forestry, environment, and meteorology. The department of environment is divided into four sub-departments, natural resources, monitoring and control, environment education, and research information.

Policy
The government has instituted a National Environment Action Plan, which is sponsored by the World Bank. This will incorporate environmental

issues in all national development plans, co-ordinate and regulate all development activities and practices which have a bearing on the quality of the environment, set policies and regulations regarding environmental protection, and advise government on the steps to be taken.

Environmental education

The government has established an institute of environmental science and natural resources at Makerere University as a long-term strategy to train manpower to promote proper management practices and utilization of natural resources. The ministry is also seriously promoting environmental awareness through radio programmes, television, newspapers and theatre arts. Many NGOs have co-operated to increase environmental awareness among Ugandans. The awareness of the problem of environment degradation has emerged and grown in many Ugandans since 1986. There is now a strong political will, and public desire, to promote the campaign against environmental destruction. There is sign of common concern.

Recommendations and conclusion

The soil and water management problems facing the smallholder farmers in Uganda, and other developing countries in general, are severe but not unsolvable. The real challenge in the alleviation of these problems lies in an integrated approach, focused on collaborative research, extension, and development. Appropriate farming technologies which take into account the farmers' meagre resources, knowledge, and preference need to be developed, to enable farmers to better manage their soil and water resources for increased and sustainable agricultural production.

The development of these technologies should consider the grassroot farmer as an active partner in the development process, in order that his or her real needs, constraints, and local knowledge are incorporated in the design of new technology. Before action is taken there should be a better understanding of how, and why, farmers organize and manage their soil and water resources in their farming systems, as well as how these systems adjust to the natural and socio-economic constraints. Such understanding will help to identify the type of applied research and extension activity that is applicable to specific circumstances of farmers.

The major problems of soil and water management in smallholder farming systems in Uganda which need to be addressed are soil erosion, low organic matter in soils, soil acidification, low soil water holding capacities, pollution of water resources, destruction of wetlands, and poor linkage between the farmer, researchers, and extensionists. The challenge is great, but with co-operation and the sharing of experiences we shall conserve soil and water resources for our prosperity.

Soil and Water Management in Zambia

N. MUKANDA and R. MWIINGA

Introduction

In Zambia, the promotion of sustainable smallholder development has focused on reducing shifting cultivation in high rainfall areas of the country. This is being tackled by a soil fertility and agroforestry research team under the Research Branch of the Department of Agriculture. In the medium rainfall areas soil conservation activities include erosion control, afforestation and agroforestry, all being promoted by the Soil Conservation Unit and the Adaptive Research Planning Team. Water harvesting, sodicity amelioration, and soil erosion research and development are the suggested programmes for the semi-arid areas of the country.

The major bottlenecks smallholder farmers face include lack of capital, credit and marketing facilities, unavailability and high cost of inputs, poor infrastructure, and poor extension services. The Zambian government has responded positively to this situation. Among the main objectives of the Fourth National Development Plan is to increase agricultural production and productivity, and to improve the living conditions of the rural population. The four specific objectives of the plan are: attaining self-sufficiency in staple foods; expansion of agricultural exports; promotion of animal draught power for agricultural production; and balancing agricultural production with environment concerns by minimizing the effect of drought, stopping soil erosion, and promoting efficient and sustainable exploitation of natural resources.

The history of soil conservation in Zambia

Organized soil conservation in Zambia started in the 1940s. Until the 1950s, soil conservation was limited to the construction of dams and weirs by the Department of Agriculture and the Department of Water Development and Irrigation. In European farming areas, such works qualified for a fifty to sixty per cent government subsidy. In the African farming areas, the soil conservation programme promoted improved farming methods which aimed at minimizing nutrient and structural deterioration of soil through the use of farmyard manure, green manure, crop rotation, and the construction of contour ridges and grass strips. Such soil conservation works on African farmlands in the tribal areas were financed by a native fund established from maize levies imposed on African maize growers. Over 1600km of contour ridges and 180km of grass strips were laid down in Southern Province between 1939 and 1944. In Eastern Province 24 000km of ridges were surveyed and pegged during 1946 to 50.

The 1950s saw a lot of activity. The Natural Resources Board (NRB) was established to supervise natural resource conservation. The actual responsibility for technical aspects of soil conservation works remained with the Department of Agriculture. The NRB encouraged farmers to form intensive conservation area (ICA) committees. The committees were administered on a voluntary basis, and were responsible for initiating, planning, and carrying out soil and water conservation works in their respective areas. Such works earned government subsidy, as well as loans obtainable by committee members from the NRB to pay towards conservation works.

In the African farming sector, the maize levy was replaced by the payment of a graduated bonus on a hectarage basis. The bonus payment was based on the area of land that was manured. But for a farmer to qualify he had to practise certain farming methods and good land husbandry, including soil erosion control.

In 1953, the government, through the Department of Agriculture started to promote soil conservation through the implementation of regional conservation plans. The previous ICAs were combined into regional planning areas in which soil and water conservation works and alignment of farm and feeder roads were constructed. By 1964, 54 regional plans covering over 1.4 million hectares in the European farming areas had been formulated. Unfortunately, farmers were reluctant to carry out farm soil conservation measures. As a result, the regional approach to soil and water conservation failed to secure the protection of farmlands from erosion.

In tribal lands, the NRB created an African Affairs Committee in 1956 to liaise with native authorities on soil conservation. But it was not until 1963 that the NRB fully extended its activities to the tribal lands. In 1964, a land use section within the Planning Branch of the Department of Agriculture was established to carry out land use surveys and planning for agricultural development schemes. The section was upgraded to the Land Use Services Division in 1966, with the responsibility of preparing regional, catchment, and farm conservation plans and the implementation of catchment conservation works. However, the responsibility for catchment conservation works was given to district councils, while that of supervising farm-level soil conservation was given to the Extension Branch of the Department of Agriculture. Although the catchment conservation planning programme expanded during the 1960s and early 1970s, it was not accompanied by the more intensive conservation measures which are necessary for the protection of farm lands from soil erosion. In this sense therefore, catchment conservation planning also failed to infuse soil conservation at farm level.

In 1970, the NRB was dissolved and The Natural Resources Conservation Act replaced the NRB with a Natural Resources Advisory Board (NRAB). Since the NRAB was to operate through part-time provincial and district natural resources committees with no technical staff, the NRAB was ineffectual. In 1972, the Department of Natural Resources was established to service the NRAB.

Research

The Agricultural Research Branch started carrying out basic and applied soil and water management research activities in the 1950s. These concentrated on water use for large-scale producers, and water harvesting and soil erosion control in the semi-dry areas of the country for small-scale farmers. The use of water conservation measures such as tie-ridging, trash farming, and ripping were tested in the semi-arid area of southern Zambia during the mid-1970s.

It is clear that soil conservation at farm level generally failed because of farmers' reluctance to implement soil conservation works. Consequently the coercive enforcement of soil conservation resulted in African farmers leaving schemes. In addition, the lack of maintenance of catchment conservation works by district councils, and the emphasis on agricultural production as opposed to conservation by agricultural extension workers, contributed to the poor performance of the catchment and farm soil conservation programme in the country.

Present status of soil and water management

The attitude inherited after independence was carried over during the period between 1970 to the mid 1980s. The same organizations still tackled conservation problems. A prominent activity during this period was that the land use branch of the Department of Agriculture was carrying out soil conservation measures on cropland only, and mainly on commercial farms. The Department of Natural Resources addressed the problems in a general way. Its main activities were directed at curing gullies and reforestation programmes.

In 1985, a programme was set up to find solutions to problems of shifting cultivation. Rapid decline in soil fertility and deterioration in the soils' physical conditions occur after one or two years of cultivation in the high rainfall areas of Zambia. In support of the wise management of natural resources, government policy, through the Fourth National Development Plan (FNDP) covering the period 1989 to 1993, has the following objectives: to arrest and reverse environmental degradation; to integrate the national management and use of human and natural resources into socio-economic development planning; to increase public awareness about environmental issues; and to ensure permanent supplies of water of acceptable quality to as many users as possible. A number of programmes and projects working towards sustainability can be cited.

The Macha Cattle Development Area Project
This started in 1982 and aims at improving pasture production and grazing management, and continuous cultivation of individual arable holdings to reduce the opening up of new crops in grazing areas.

The National Soil Conservation and Agroforestry Extension
This programme fosters soil conservation on arable land, grazing land and homesteads, stream banks, and gullies. Tree planting for soil fertility improvement, cattle feeding, and for wood fuel is a major component of the programme.

The Community Forest Programme
This encourages local communities in wood-deficit areas to set up woodlots for fuel and other purposes, and consequently to reclaim degraded environments.

Luangwa Valley Integrated Resources Development Project
The goal is integrated rural development with resources conservation to provide a self-sustaining future for both the human and wildlife populations.

The decentralization of the Natural Conservation Strategy

Under this programme, the Department of Natural Resources is reorganizing provincial and district natural resources committees into cross-sectoral institutions with the ability to supervise and manage the utilization of natural resources.

The Department of Natural Resources continually monitors the status of environmental degradation outside gazetted forest estates, national parks, and fisheries. These are areas which are mostly occupied by smallholders. In catchment areas, early burning is conducted to enhance soil and water conservation. In other areas, curing of gullies, tree planting in deforested areas and other appropriate rehabilitative measures are done.

Collaboration

A number of international agencies are currently collaborating in projects. The International Board for Soil Research and Management (IBSRAM) is involved in the Zambian Soil Productivity Research Programme. This programme carries out studies in agroforestry, soil acidity and aluminium toxicity, nutrient dynamics, trace element deficiencies, soil erosion, and farming systems. Zambia is a member of IBSRAM networks on tropical land clearing for sustainable agriculture and on management of acid tropical soils. The Soil Productivity Research Programme also collaborates with the Tropical Soils Biology Fund in soil micro-biology studies.

Conclusion

The Zambian Research Action Plan adopted in 1990 demands that soil mapping and crop suitability evaluation should be carried out in all areas where major research activities are to be conducted. This activity will ensure a better understanding of crop and soil water management and conservation requirements, and soil fertility improvement and agroforestry needs. Poor rainfall distribution in recent years suggests need for an emphasis on water management. The high cost of inorganic fertilizers is casting doom on the small-scale farmers. Thus the incorporation of tree shrubs in farming, as well as in cropping systems, as sources of nitrogen is important.

Soil and Water Conservation in Zimbabwe

S. KAVALO and G. NEHANDA

Introduction
Zimbabwe is a land-locked country covering some 390 000km^2, apportioned into the following categories: communal areas, 163 600km^2; resettlement farming land 26 400km^2; commercial farming land 142 400km^2; national parks estate 47 000km^2; state forest land 9000km^2; and urban and state land 2200km^2. The population is about 8.5 million, and approximately eighty per cent of these people are poor and live in the communal areas.

History of soil conservation

Pre-colonial period
During the pre-colonial period the population was very low and people practised shifting cultivation. Plots were small because people used hand implements. Abandoned plots were left to fallow and little environmental damage occurred.

Colonial period
The 1890s brought an influx of Europeans whose main objective was to search for minerals, especially gold. The prospect of finding gold proved to be difficult, and these people had to be compensated by land. Huge chunks of fertile land were given to European settlers by evicting indigenes and forcing them into reserves (communal areas), where the soil was less fertile and the rainfall low. As both human and livestock populations grew, land was no longer sufficient for people to practise shifting cultivation. They started to adopt a settled type of agriculture. With increased population pressure came land degradation through overstocking and overcropping. Consequently, soil erosion became a serious problem. However, measures to curtail the problem were not introduced.

Between 1913 and 1926 environmental problems were becoming apparent. The population had risen to 936 000, and the colonial government was compelled to draw up conservation policies to correct the situation. Two major laws pertained to the construction of contours in crop lands and destocking of rangelands. These failed because they were restricted to commercial farming areas only. In 1939 the state of soil erosion and destruction of natural resources in the native reserves was serious and needed vigorous schemes to avert disaster.

In 1941 the Natural Resources Act was promulgated. This called for the conservation and improvement of natural resources, and for the creation of a Natural Resources Board, which was later established. The board was to be the public trustee of the natural resources, and its roles were to raise public awareness, to supervise the usage of all natural resources in the country, and to recommend appropriate legislation to the government.

Local community participation in conservation matters was recommended and the Natural Resources Act allowed for the creation of conservation committees in farming areas, a feature which remains to this day. Most of the activities of the board and its voluntary conservation committees were confined to commercial farming areas. Although there were similar committees in the reserves, they were not as effective. In view of this situation it was felt that agricultural reforms could only be effective if they were made compulsory. The legislation to carry out such reforms was formulated in the Native Land Husbandry Act of 1951, and was implemented during the period 1956 to 1962. There was no attempt to educate the indigenous poeple on land management. Conservation measures were imposed on the peasant farmers.

During the late 1960s, and through the 1970s, the civil war took its toll on the peasant farming communities and the possibilities of improving their agriculture were reduced. These problems were further compounded by the increase in population of livestock and humans, and extensive land degradation resulted.

Post-colonial period to the present
The new government inherited a dual agricultural system. On the one hand there was the advanced commercial farming sector dominated by large-scale farms and showing all signs of environmental awareness. This was largely because all extension services were confined to it through a body known as Conex (Department of Conservation and Extension). On the other hand was the communal land/peasant sector, characterized by problems of population pressure and land degradation. Land tenure systems were also different. The former sector being dominated by freehold title, while in the peasant sector communal use of land prevailed.

In trying to correct imbalances the government had to embark on massive policy and programme changes. It instituted a massive programme on conservation education and extension and developed the concept of the conservation movement, as detailed in the Natural Resources Act, to include communal areas.

Priorities and strategies

The government's priority is for the construction of standard mechanical conservation works in every cropland. Each farmer has to put in some mechanical conservation work before tilling the land. Once an individual is allocated land he or she must approach an Agritex worker who will place the farmer on a name list for the pegging team. To ensure compliance natural resources officers make regular inspections, to check on defaulters and those who have sub-standard conservation measures. Section 75 of the Natural Resources Act, chapter 125, also gives powers to natural resources officers to stop the farmers from further cultivation until the prescribed recommendations are followed satisfactorily. People have responded positively to this message and are coming forward requesting for conservation works to be pegged in the arable lands.

In the resettlement areas, there is no provision for the Department of Natural Resources and the Natural Resources Board to move in and form conservation committees for promoting conservation awareness. Where a resettlement officer is conservation-conscious and co-operative, environmental problems are reasonably under control.

The education programmes have increased awareness of environmental problems. People are now starting to propose their own conservation projects, participating fully in consolidated gardens, gully reclamation, improved grazing schemes, rural afforestation, and water harvesting.

Challenges

There are three main groups of agricultural land use that are in extensive use in Zimbabwe today:

- The commercial farming sector (thirty-six per cent of the country)
- Resettlement sector (seven per cent of the country), and
- The communal sector (forty-two per cent of the country) land groups.

The commercial farming sector is dominated by highly skilled farmers who have access to resources and can afford to adopt sophisticated technologies in farming. In general terms, ecological balances have been maintained on these lands through good conservation management. High levels of production efficiency are achieved, with crops contributing some eighty-five per cent of the gross marketed agricultural product.

The communal lands are the home of poor subsistence farmers with limited access to resources. Their methods of farming are predominantly exploitative in nature and have a negative impact on the ecological balance. Nearly ninety-one per cent of the communal lands occur in agro-ecological zones III, IV and V, which are marginal in terms of agricultural productivity. These are characterized by low rainfall with poor distribution, short season length, and low soil fertility. It is on these lands that over eighty per cent of the Zimbabwean population derive their livelihood. There are immense land and water management problems associated with these lands, such as:

- Sheet erosion in the communal areas has been estimated at 50t/ha/yr and rainfall loss as surface runoff to be about thirty per cent of total seasonal rainfall
- The current agricultural practices employed in the communal areas result in untimely planting, rapid decline in soil conditions and available depth
- If soil and water management problems are allowed to continue unchecked, the economic base of the communal area will be destroyed. This will plunge the rural people into deeper poverty, and force the government to seek alternative lands for resettlement and divert much needed development resources to feed the poor, and
- The bulk of soils in the communal areas are sandy soils derived from acidic parent materials. Their low fertility status has been worsened by traditional farming methods that tend to degrade the land resource base

through continual cropping with little or no organic or inorganic fertilizer inputs.

The smallholder farmer lacks both financial and material resources to build up and maintain a high level of fertility on his or her land. The potential productivity has been declining steadily over the years and is now threatening to put large tracts of land out of production.

National programmes

Policies and strategies

Punitive laws have been ineffective in correcting environmental degradation. The department's new approach includes a massive education campaign. Its aims are:

- To inform the communities about the danger of reckless utilization of their resources
- To explain the importance of sustainable utilization of their resources
- To impart proper land husbandry practices, and entice the communities to adopt them, and
- Initiate conservation projects in association with the education campaign.

To improve effectiveness, the department has established a research and training branch. This identifies new scientific technologies, ideologies, and information which can be adopted. First, the research and training sector trains the natural resources officers, who in turn organize workshops and awareness meetings to impart the new ideas to the farmers.

There has been a shift in national policy since 1980 to give more service to the communal areas through increased research and extension work. The government's main goal is to ensure the integration of sustainable resource use with every aspect of the nation's socio-economic development. Research programmes now cater for the communal area needs in all agro-ecological regions.

Water conservation and management

In the Lowveld of Zimbabwe, the major constraint to crop production is lack of water, so research has been started to devise water harvesting techniques. To maximize the use of rainfall, the tied-furrow system has been employed, and is reported in the paper by Nyamudeza *et al.* in this volume.

Research and extension linkages

The beginning of on-farm research marked a new era in the co-operation between research and extension agencies. An on-farm research planning committee comprised of DR&SS (Department of Research and Specialist Services) and Agritex was established to assess relevance of proposed DR&SS trials and demonstrations and Agritex recommendations, to define appropriate site selection procedures given the wide diversity of communal area ecologies, and plans to monitor the implementation of on-farm trials.

Conclusion

The government would benefit from the introduction of an overall policy to direct the management of all natural resources. Steps to formulate this policy were initiated by the government in 1983 by drafting a national conservation strategy. Policy guidelines have been drawn for the desired course of action of the management of all natural resources. The population factor, mass participation, and education have also been included. The success of the strategy, once it becomes policy, is yet to be seen. This would depend on the preparatory ground work which the people and government would have embarked on prior to its implementation.

Zimbabwe is currently developing systems for sustainable agricultural production in the smallholder sector. It has generally been realized that for permanent solutions to the land and water management problems in the smallholder sector to be established, socio-economic factors must be incorporated in the development of appropriate technologies. Diagnostic surveys are now being adopted as a way of identifying research programmes and areas of demonstration. These methodologies are now being emphasized both in research and extension.

Current Aid Agency Approaches to Soil and Water Conservation

Compiled from papers by
P. EWELL, D. HUGHES, D.W. SANDERS and R. GALLAGHER,
J.K. RANSOM, A. WOOD, C.S. WORTMANN

Introduction

Soil and water conservation (SWC) has undergone a transformation during the past decade. This paper draws together the experience of several agencies and identifies the main elements in the new approaches. It provides a point of departure from which progress can be made towards developing SWC approaches which are especially suitable for smallholders.

The agencies considered here have different approaches to SWC. Some see it as an issue which has to be addressed directly. Others are more concerned with raising the welfare of farmers and the productivity of their land use systems, thereby helping improve SWC indirectly.

FAO approaches to soil conservation and rehabilitation

The Land and Water Development Division of FAO has recently been involved in a major review of its experience in SWC. From this a number of conclusions have been drawn (FAO, 1990a, 1990b). FAO believes that SWC must change from top-down attempts to treat land degradation in isolation, to an approach which is participatory and sees SWC as part of improved land use. FAO also recognizes the complexity of the processes influencing soil erosion. It understands that a more holistic approach is needed which involves interventions and changes in a wide range of socio-economic, as well as agronomic and environmental issues and the consequent policy adjustments where required. FAO is committed to the concept of sustainable productivity through a user-driven approach to problem solving, with decreasing reliance on mechanical conservation measures.

FAO believes that SWC will be more effective if three particular areas receive attention. These are: improved land use; land user participation; and the institutional development.

Development of appropriate systems of land use is important because well-managed crops, pastures, and forest, not only offer protection and rehabilitation of the soil resources, but link conservation and rehabilitation to increased yields and better production. This is vital to achieve the support of farmers as it makes conservation and rehabilitation attractive to land users and produces immediate benefits for them.

Land users have to be convinced of the value of changes in their current systems before they will adopt any new proposal. Hence, soil conservation measures are more likely to succeed when low-input techniques are proposed which are both practical and socio-economically viable. To ensure this it is important to involve land users in the identification of low-input and productivity-sustainable solutions, while helping them adapt many of their customary practices.

FAO recognizes that there is an important role for the state, and not just the farmer, in addressing the causes of soil erosion. Often the underlying problems involve issues of land tenure, the pricing of agricultural crops, subsidies and incentives, and social customs and legislation, all of which require the state's attention. Because these have not been considered as conservation issues in the past, many SWC programmes have dealt with the symptoms, rather than the causes, of problems. As a result, the impacts of projects have been short-lived.

A government's main function, according to FAO, must be to create conditions which encourage land users to practise sustainable and more productive forms of land use, rather than enforce the adoption of specific soil conservation measures. However, there is no single solution to soil conservation, and the actions required of government may be in a variety of economic and socio-political policy areas, as well as in extension and research to produce economically attractive combinations of land use that also address the need for soil conservation.

Governments who, after all, have the prime responsibility for halting continuing degradation, should formulate appropriate plans and implement projects, and mobilize people to improve participation in problem identification and solution. This will involve making people aware of SWC problems and encouraging them to practise new land use management.

This can be helped by governments encouraging the formation of land user organizations or associations, to assist with demonstrations linked to publicity about land productivity and conservation issues.

Besides local institutions FAO supports the development of national institutions, to provide the necessary back-up services. Establishment of a high-level advisory committee is also recommended to formulate conservation strategies and develop policy co-ordination and monitoring aspects linking SWC to national budget and policy planning departments.

IUCN – The World Conservation Union

IUCN links governments and NGOs in efforts to conserve bio-diversity, maintain natural ecological processes, and achieve sustainable use of natural resources. These are all seen as elements of ecological sustainability, which is necessary to ensure that the needs of present populations can be met while protecting the ecological heritage of future generations.

IUCN recognizes the complexity of the SWC problem and the need for interventions, both at the field level through projects, and at national level through policy co-ordination and the integration of ecological issues into national development planning. IUCN also stresses the need to work with governments and community institutions to help them develop the capacity to address the complex problems which are faced in SWC.

IUCN field projects typically have a long time horizon in order to ensure that the dynamics of natural resource use, and the perspectives of the natural resource users, are fully understood by those involved with the management of projects. Besides a long-term commitment, IUCN seeks a high level of participation and partnership by the local community and authorities. This provides the basis for the participatory research which is needed to find appropriate solutions and innovations. A long-term partnership is also needed in order to develop the skills and capacities within communities and local administrations in order to bring local knowledge to bear, and for the project population to understand the role and use of specific natural resources within local environmental and economic contexts.

In east Africa IUCN has two projects which are addressing SWC in an integrated manner, and two with an SWC component (the Serengeti Project in Tanzania, and the Kibale-Semlike Conservation and Development Project in Zaire). The integrated projects are the Mount Elgon Conservation and Development Project in Uganda, and the East Usambaras Agricultural Development and Environmental Conservation Project in Tanzania. Both projects are concerned with the management of mountainous areas where serious erosion and increased runoff are occurring as a result of deforestation and changes in land use and land management. These changes have been caused by a variety of factors, including uncontrolled timber extraction, fuelwood collection, grazing and agricultural expansion into the formerly forested areas, changes in farming systems, and population growth.

The overriding objective of these projects is to bring deforestation under control, because forestry offers the best sustainable use of the upper terrain of these mountainous areas. In both projects this goal has been formulated by the local community after a process of discussion and exploration of the issues involved.

A key element in the projects is understanding the pressures upon the local communities which lead to encroachment into the forests. Once these are identified the local communities are supported in developing adjustments and innovations in their farming systems and rural economy. These improve standards of living in a sustainable manner within the areas already settled, and through controlled and sustainable use of some of the forested areas. A key aspect of the East Usambaras project has been the appointment of community-based extension officers, or village co-ordinators, who help communities prioritize problems and undertake participatory research to develop appropriate solutions. The Mount Elgon project works through the existing government district and parish extension structure, which has much the same end result.

Innovations within these projects have included land intensification with minimum inputs, through innovations which help improve the productivity of labour and so make intensification more attractive. As a result there has been an emphasis upon low-cost, locally-resourced changes, with agronomic soil and water conservation measures, crop additions and agroforestry methods of improving soil fertility, intensifying land use and reducing the need for grazing land. Other innovations have been the development of fish ponds.

These projects have developed their understanding of the problem of forest clearance over the past five years. Initially, the focus was on controlling clearance of the forest through the development of a buffer zone at the forest edge. However, it has been increasingly recognized that this has to be seen in a broader context, and the emphasis is now upon the participatory development of community-wide natural resource management plans, which include the community having a say over use and management of the forested land. This has led to a broadening of the project, so that the project covers not only the forest fringe but the whole area which the community uses, and the whole range of activities by which the community live.

Whatever the initiatives at the local level, they may founder if the overall milieu in which they are proposed is unfavourable. Hence IUCN supports national initiatives to develop policies and institutional sensitivities which will create a favourable environment for local level initiatives. One aspect of this is the development of national conservation strategies, but other actions may include the development of policies for specific areas or types of land.

CARE and agroforestry

CARE is an NGO primarily concerned with improving the income, health and nutritional status of people through the attainment of sustainable productive ecosystems. CARE recognizes that environmental sustainability and human development are mutually dependent, and believes that environmental degradation is a major cause of many natural disasters and underdevelopment.

Since 1974 CARE has developed 16 projects to address natural resource management problems in east and southern Africa, taking agroforestry as the core element in helping families meet their food and timber requirements. The project designs have involved offering communities a wide range of tree planting configurations, using techniques such as terracing, infiltration ditches, and hedgerows to assist local people improve soil conservation and fertility. Of these 16 projects only one has the specific aim of reversing soil erosion. However, improved soil conservation and fertility are anticipated in most of the projects.

A basis for all of CARE's projects is a recognition of the link between an individual's perception of the costs and benefits of change through adopting an innovation or development activity and his or her willingness to expend personal resources on that. Nurseries flourish where they are able to harness communal inputs of water and labour and transform them into a product to be marketed on its own merits. Political support from local chiefs and government officials is also important to achieving a favourable reception to initial ideas for interventions. Although it is the specific benefits for farmers which have determined the success of projects.

The Siaya and South Nyanza Agroforestry and Extension Project was developed in 1984 in an attempt to address the problems of declining fuelwood on public lands and reduced fallowing due to declining farm size. During the first four years of the project nine million trees were planted using 23 different agroforestry interventions and 36 different tree species. Techniques used included alley cropping, woodlots, fodder cropping, border planting, fruit orchards, windbreak systems, live fences and hedgerows. Through the provision of a choice of agroforestry techniques CARE hopes that farmers can identify innovations which are relevant to their circumstances and provide attractive returns, not only in the form of fuelwood, charcoal, and fodder but also in the form of improved soil conditions and agricultural production. The project is based on a participatory extension technique through which the community and CARE can build upon indigenous techniques of integrated crop and tree production. In this project soil and water conservation is an indirect benefit for farmers, the primary gains being in fuelwood and soil fertility maintenance. In this way SWC is adopted because it offers clear benefits in terms of improved production.

On the island of Anjouan in the Comoros Islands, population growth and the decline of a spice-based cash crop economy are leading to increased pressure upon the steep land of the island with serious implications for SWC. While there is a clear need to stabilize soil, farmers are most interested in improving production under these conditions of land shortage. The technical package promoted was initially vetiver grass (*Vetiveria spp*) planted along contour lines with occasional trees, usually filao (*Casuarina equisitifolia*). The grass inhibits downward movement of soil, causing gradual soil accumulation and passive formation of terraces. The trees provide deep-rooted anchorage for the terrace faces. The resulting terraces have gentler slopes which reduce soil erosion and stabilize the land. Their main benefit from the farmer's viewpoint is not soil conservation, but the way this makes investment in soil nutrient restoration more attractive. One adjustment which has occurred during the project has been the replacement of filao by *gliricidia*, which can be used as a fodder source, while vetiver has been replaced by Guatemala grass (*Tripsacum laxum*) and *Pennisetum purpureum*, both of which are more palatable fodder species. In this way the returns to farmers are improved and SWC is made more attractive, even though it is not the farmers' main goal.

CARE agroforestry projects involve the use of different techniques to achieve the same goals of improved welfare through increased food and wood production on farmers' smallholdings. It is recognized that farmers must be presented with a choice of innovations so they can identify those which are best suited to both local environmental conditions and also to their personal circumstances. This requires a participatory approach so that potential external innovations are properly adjusted to local conditions, or developments can be made to existing indigenous practices. Further it is seen that farmers' priorities are meeting their immediate needs rather than long-term environmental conservation. Thus CARE projects focus on meeting those specific needs first, although improved environmental sustainability through better SWC is seen as an essential part of all CARE projects.

CGIAR and the role of improved returns

The CGIAR centres have specific responsibilities for long-term basic research on breeding and selection in order to identify improved genetic material for testing and further development by national research centres. This they seek to do by strengthening the capacity of national programmes through collaborative research, training, consultative services, and networking activities.

Because improved production is seen as the overriding goal, the CGIAR centres do not have a specific mandate for soil and water conservation. However, their work makes an important contribution to SWC because of the way in which the centres seek to identify innovations which produce improved plant growth and more efficient nutrient use. In this way SWC is made attractive for farmers, and is achieved as a by-product of production enhancing innovations.

The international centres work with national programmes on two different types of technology. The first group is production technology which broadly refers to the methods which farmers, processors and middlemen use to produce, store, transform and market their products. Specific technologies include maintenance of genetic diversity, integrated pest management, and tissue culture for rapid multiplication. The second group is research and development technologies which refer to the methods and organizational strategies used by researchers and technology transfer institutions. These have typically included diagnosis and priority setting, farmer-back-to-farmer forms of participation, post-harvest management, soil fertility management, co-ordination and management of on-station and on-farm research, research-extension linkages, monitoring and evaluation and collaborative research networks. The application of these production and R & D technologies are important for SWC today in several ways as will be clear from the experience of the different agencies.

CIAT

The International Centre for Tropical Agriculture (CIAT) has a global mandate for bean research. It has established three regional programmes in Africa to work in collaboration with national bean research programmes and other institutions. Beans are important in eastern and southern Africa, especially in the highlands of east Africa where annual per capita consumption typically exceeds 50kg. Many of the countries in this region also suffer serious soil nutrient depletion. Germplasm preservation and enhancement and the development of crop and soil management systems for increased, but sustainable, productivity are the two major goals of these programmes. CIAT is involved with improving the efficiency of nutrient use and conserving and supplying nutrients. But this interest is not restricted to soil erosion problems, and includes losses of nutrients by leaching, volatization and exportation.

Aspects of the research programme which are relevant to SWC include:

- Enhancement of biological nitrogen-fixing, which will increase yields and enhance plant growth and increase ground cover
- Improved adaptation to edaphic stresses, such as low nitrogen availability, low phosphorus availability, aluminium toxicity and manganese toxicity. This can be achieved by genetic selection of varieties with improved nutrient use efficiency, or in the case of nitrogen improved biological nitrogen-fixing. This will raise yields and increase ground cover, while genetic improvement can enhance the ability to extract phosphorus of low availability and put it into a more recyclable form, and
- Improved crop and soil management systems with better sustainability, as most traditional systems of management are exploitative, with rapidly declining resource bases. To develop such systems requires: study of farmers' systems and objectives; evaluation of alternative management technologies; and improved application of existing information.

The central focus is the search for technologies which complement or add value to nutrient conservation measures to make them more attractive. The ability of beans, either in sole crop or intercrop, to establish early ground cover is important for erosion control on sloping lands, and fortunately, there is a positive relationship in beans between early vegetative growth and yield (Wortmann, 1991). By selecting bean cultivars for increased yield potential, bean breeders are also selecting cultivars that give earlier ground cover.

Some yield-increasing technologies under evaluation by bean researchers have the potential to enhance the desirability of certain conservation practices. The adoption of climbing beans in areas where they are not traditionally grown is creating a demand for stakes which can be supplied from strips of *Pennisetum purpureum* or hedgerows installed for erosion control.

Crop soil interactions are important for nutrient use efficiency, crop growth, and erosion control. Innovations in these areas must be identified and evaluated through participatory methods involving both farmers and extensionists. While farmers can fine-tune technologies, these must be adequately adapted in the first place to have obvious benefits for farmers.

CIMMYT

The International Maize and Wheat Improvement Centre (CIMMYT) defines its current mission as 'to help the poor of developing countries by increasing the productivity of resources committed to maize, wheat, and triticale, whether in research or on the farm' (CIMMYT, 1989). Small-scale farmers are often resource-poor with limited access to cash, credit, inputs, and labour, which prevents them producing their full economical yield potential.

CIMMYT's research is based on:

- Realization that farmers often operate complex production systems, have multiple objectives and make rational decisions
- Careful identification of farmer clients, as recommendations have to be targeted to specific groups
- Recognition of technical/economic interactions within the family farm which affect responses to alternative technologies, that is technical excellence may be compromised in the interests of productivity of the whole system
- Involving both biological and social scientists in the research process, and
- Carrying out much of the research on the farms of representative clients under their productive systems.

Based on these elements CIMMYT has developed a research procedure known as on-farm research with farming systems perspectives (OFR/FSP). This requires researchers to be sensitive to farming systems interactions and to farmers' viewpoints. Understanding these must be central in the design and evaluation of research activities. Thus the key to CIMMYT's philosophy of developing technology for the small farmer is the need to understand all the farmer's circumstances, physical, biological, social, and economic.

While CIMMYT does not have any research directed specifically towards soil and water conservation, this is included in sustainability which is one of the 12 interrelated criteria used to establish CIMMYT's priorities towards the twenty-first century (CIMMYT, 1989). (Six of the others were related to efficiency, four rest on concern for the poor, and one relates to sub-Saharan Africa.) One particular area of concern to CIMMYT is the expansion of agricultural production into more marginal areas where there are major problems with moisture shortage, temperature extremes, nutrient deficiencies and toxicity, insect pests and diseases. SWC and more efficient use of water and soil nutrients is one element of the research for sustainable agriculture.

Of the CIMMYT programme activities directed towards sustained increased productivity, germplasm development is seen as one way of achieving greater efficiency in the use of water and soil nutrients and so increasing yields. However, there is an indirect contribution to SWC because, by producing better plants, greater ground cover and protection is provided for the soil.

The CIMMYT crop management research programme is problem-oriented and uses on-farm research. This has led to a number of research programmes concerned with land degradation as a problem faced in maintaining productivity. In the east and southern Africa region research on such topics includes zero and minimum tillage systems for maize, and control of the weed *striga* in maize-based systems.

Finally, it should be stressed that CIMMYT recognizes that most sustainability problems, including SWC, are complex and cannot be solved by technology alone. Institutional development, and sensitivity, coupled with appropriate policies are needed in many of the complex situations which are faced if sustainability is to be achieved. CIMMYT is paying increasing attention to sensitizing national programme agronomists and economists to major long-term sustainability issues.

CIP

The International Potato Centre (CIP) has a mandate in east and southern Africa to support research on Irish and sweet potatoes. The former (*Solanum tuberosum*) are grown at high altitudes on steep slopes, and are an important food security crop, as well as a cash crop for urban markets. In these areas farms are small and population pressure is high. The potato is important in the rural economy because of its high calorie yields and the cash income it generates, while it is ecologically important because of the ground cover which it provides. Sweet potatoes (*Ipomomea batatas*) are grown in the densely settled mid-altitude areas where they are an essential part of the starch staple diet. They are increasingly grown in semi-arid regions where the population is expanding rapidly. Here despite marginal and variable conditions they are an important source of household food security in periods of the year when other foods run short, and also in drought years. In these changing conditions a dynamic approach is needed for the development of new technologies.

The main focus of CIP work is on:

- Genetic research to increase the diversity of varieties appropriate for a range of conditions, and which are resistant or tolerant to pests and diseases
- Interdisciplinary research which seeks to minimize inputs and other resource costs

- Diagnostic studies of the changing conditions and needs in specific agro-ecological zones
- Institutional development of seed systems and other mechanisms required to meet farmers' needs, and
- Research on soil fertility, pesticide management and other specific aspects of sustainable technology.

For over a decade CIP has emphasized the importance of organizing research in response to the needs of farmers, consumers and other potential users of technology. The farmer-back-to-farmer model synthesizes two important concepts which are now widely accepted. The first is that farmers should participate from the first stage of research, the diagnosis of the problem, through the processes of technology testing, adaptation, and monitoring. The second is that an interdisciplinary team approach is the most effective strategy for research and development of technology for resource-poor farmers.

Donor Perspectives

JOHN LYNAM

Interests and roles of donors

A significant number of donors provide funding for various aspects of soil and water management in east, central, and southern Africa, and some have been active for almost 20 years. Funding for this area has been given added impetus by the recent support for sustainable development. One of the highest priorities on the sustainability agenda is the maintenance of the productive capacity of the land resource base. It is recognized that resolution of the multifarious land use problems in the region will require a continued long-term commitment by donors in soil and water management.

The role of donors revolves around the central function of funding. Funding for soil and water management is defined along two or three principal dimensions. In the institutional dimension, funding may be provided to international organizations, regional co-ordinating bodies such as SADCC, or networks, national ministries or parastatals, or NGOs. Funding is provided across the continuum from basic research, through applied and adaptive research, to development activities, encompassing a range of options such as community development projects or watershed management projects. There may be a country focus, an agro-ecological zone focus, or a simple problem focus, such as the management of vertisols. The power of funding is that it can shift priorities across these various dimensions and effect improved integration and organization of activities.

The research and development continuum

The relative priorities in soil and water management across the continuum are not clear. New avenues of research in the management of tropical soils are opening up. Our knowledge of how to link improved soil management techniques to farmer needs in Africa is only in its infancy. Each new research area has required a focused, specialized programme. Thus, ICRAF works on agroforestry, the Tropical Soils Biology Fund (TSBF) on the purer research aspects of soil organic matter management, the University of Nairobi on soil erosion, IBSRAM on acid soils and vertisol management, NIFTAL on nitrogen fixation, and IFDC on chemical fertilizer use and plant nutrition. Donors provide funds for all these activities, as there is no one solution to maintaining productive soils and is no clear basis for establishing priorities across these areas.

Priority setting will only come with more experience at the farm level in evaluating adoption of management practices. How to select a management package that will be most appropriate for any particular farm situation remains a clear challenge. Thus, donors are interested in funding new approaches to on-farm technology improvement in the soil management area, such as the identification of effective participatory research and development methods. New ground is being broken by innovative work, but much remains to be done. A body of documented experience will be needed on what motivates farmer investment in new soil management practices, the conditions under which these are profitable and fit within the farmers' resource constraints, and the extent to which these are driven by market versus subsistence objectives.

Donor commitment and priority setting

Donors recognize the need for a long-term commitment to funding activities in soil and water management, and that funding has to cover the spectrum of activities from basic research to development projects. Yet this recognition raises the question of whether it is possible to achieve a better integration and organization of these activities, especially given the diversity of research institutes, donors, and development agencies operating in the area of soil and water management.

The obvious question is, who should establish funding priorities? A wider donor–client participation in the setting of priorities has occurred at the national level, and the prime example of such efforts are environmental action plans. These are prepared to accord with country priorities as determined by some form of limited consensus, primarily across government agencies. The intention is that donors fund components of the plan based on their respective programme focus or expertise. Getting more widespread participation in priority setting will require a lot more donor co-ordination, increased flexibility, and less control in donor programming of funds.

Are there modalities for better donor co-ordination? Environmental action plans may help to co-ordinate funding around a particular topic in a particular country, but they do not change the way different donors structure their funding programmes in-house. Co-ordination will be difficult to achieve because donors need to establish their own identity, primarily to garner political support for funds from their respective governments. Funding agencies must often walk a thin line between accommodating the political fads in the home country, and trying to develop a long-term funding programme for critical areas in Africa, such as soil and water management.

Priority setting at the country level usually involves government policy and planning. Now there are new moves to fund NGOs, in the hope of targeting benefits directly at local people. Because of the crisis in public finance and administration within African government agencies, donors are increasingly looking to NGOs as a vehicle for providing services to farmers, especially in more development-oriented projects. Some donors use international NGOs as intermediaries to link with and fund local NGOs. Other donors fund local NGOs directly. There has not been much experience with this approach to funding, and it is impossible to say at this time whether increased funding channelled through NGOs will be effective. Moreover, it must be recognized that such shifts should not reach the point where there is a clear competition between NGOs and government ministries for donor funds. Complementarity and collaboration must be the keys to successful expansion of NGO programmes.

Should the farmer or community have an input into donor decision-making through participatory relationships? The probability is that the donor will very rarely interact directly with community organizations. Programmes will mostly be wholesaled through intermediary organizations, be they NGOs or government agencies. Administrative costs are just too high for donors to develop linkages with many small organizations, not to mention the sovereignty issue as regards bilateral donors. Nevertheless, donors are concerned about whether farmers needs are being addressed in this wholesaling process. There is an increasing trend by donors to build monitoring and impact assessment capacity directly into projects. This, of course, still does not ensure that community views are incorporated into the project design. Working out formalized contracts for project development and implementation between the community and the intermediary will guarantee interaction, and would seem to be the way forward.

Socio-cultural Issues

Introduction

EVA TOBISSON

Insufficient attention is paid to socio-cultural issues, to the potential of rural sociologists and anthropologists, and to farmers' participation in development activities.

The significance of the human dimension in natural resource management can no longer be neglected. The materials in this section state some of the rationale for this and show, by example, how rural sociologists have contributed to development. One paper describes some of the critical fields where a sociological input is desirable. Another contribution reports work from Zambia, and is of particular interest because it demonstrates partial integration of sociologists into a branch of a line ministry, some of the hurdles to be overcome in developing rural sociology, and some sociological successes.

Very few countries have established roles for rural sociologists in the mainstream of development activity. The points are made that there are major roles, and sociological contributions are needed in the context of an inter-disciplinary approach to sustainable development. For participatory development sociological input will be necessary if the dynamics of family and community are to be understood, so that improved land husbandry practices can be matched to satisfy peoples' needs.

Changing Roles for Rural Sociologists

EVA TOBISSON

A change in perspective

A specialized natural science is the conventional view of soil and water management. This is reflected in the kind of expertise assigned to projects and agencies, in the formulation of policies and strategies, and in measures put forward to deal with problems. In recent years we have witnessed changes towards a more comprehensive approach, in which the complex relationships between local people and their natural resources are given attention and new roles for local inhabitants are identified. Local people used to be perceived as the principal cause of environmental degradation, now they are increasingly looked upon as a powerful resource for action to improve land husbandry. Recent experiences have shown that local people, and poor people in particular, have a vested interest in environmental protection because their livelihood is dependent on careful use of natural resources.

The change in perspective has mainly been inspired by two important insights concerning the shortcomings of conventional project approach to natural resource management. First, projects have invariably been designed and launched by outsiders, while the role of the intended beneficiaries has been confined to that of passive receivers of project benefits (Paul, 1987). Second, projects have been planned and implemented without sufficient knowledge of community features which have a bearing on natural resource management, to the effect that neither the potentials nor the constraints inherent in the community have been given recognition. The lack of participatory approaches has worked against the objective of sustainable resource management. If development is about people, projects and programmes must be built around their needs and capacities to take action.

Popular participation

It has become increasingly clear that past efforts to meet the specific natural resource management and general development objectives have failed for want of participatory approaches. The notion of participation has in itself constituted a problem. When participation has been on the agenda its meaning has generally been confined to local inhabitants contributing their labour and other local resources. This restricted interpretation has been common in relation to soil and water management projects since the restorative measures proposed have tended to be excessively labour demanding, for example dug terraces. The interpretation is changing from a narrow focus on involvement in the implementation to participation in problem identification, planning and design, as well as in the monitoring and evaluation of project activities and results. Experiences have shown that project activities stand a greater chance of being sustained when the intended project beneficiaries are invited to participate throughout the project cycle, and are granted decision-making power.

Communities differ greatly with respect to a range of factors, such as ethnicity, agro-ecological conditions, previous experience of external intervention, degree of integration into the market economy and proximity to urban centres. Moreover, a community may be treated administratively as a homogeneous unit of people who share the same needs, aspirations and perspectives, when it is in reality a complex entity exhibiting major differences and contrasts between the constituent groups. For analytical and operational purposes, project models have to be adapted to local conditions and the community has to be broken down into groups (based on gender, age, economic status, and so on) that can be expected to have common problems, responsibilities, rights, value systems, interests and opportunities for change.

The change towards a people-oriented focus necessitates a shift to an interdisciplinary approach, in which social science expertise dealing with micro-level socio-cultural and economic aspects of resource utilization and management should be incorporated.

Anthropological and sociological aspects of land husbandry

Research dealing with developing countries is concerned mainly with social stratification, gender relations, people's management of ecological and human resources, changes in land-use and land-management, and the consequences of local communities being absorbed into wider economic and political systems. The aim is to understand and explain societies through analysis of how they are perceived by the people within those societies. Although this understanding is mainly obtained through long-term field studies, short-cut methods have been developed.

Anthropologists and sociologists have an important role in relation to natural resources management through their work to understand the functions and structure of local communities. For example they can help define target groups, and predict unintended consequences of planned change. They can help detect local expertise and acknowledged community leaders who could expedite participatory development.

Costs and benefits from the perspective of local people

Economic analysis and arguments in relation to environmental degradation have become increasingly influential over the last few years. Cost-effectiveness of environmental management activities is important, but anthropologists and sociologists are concerned that economic considerations are based on neo-classical economics, where the assumptions are about individual preferences and profit maximization. These often carry greater weight for there is rarely a costs and benefits analysis from a community perspective.

Peasant economies do not operate according to the economic laws assumed by neo-classical economists. For example, risk minimization and family subsistence, rather than profit maximization, constitutes a fundamental principle. Therefore, sound economic analysis must be broader than the conventional cost-benefit analysis. Cost-benefit analysis about groups brings us into a complex web of socio-cultural, political and economic considerations where a sociological input is necessary.

Tenurial rights

Indigenous tenurial rights are complex and not easily understood. Any project aiming at sustained natural resource management must be based on a comprehensive understanding of those rights. This is another area in which anthropology and sociology can make an important contribution.

A condition for sustained natural resource management is that local groups and individuals have secure access to land and to other natural resources required for livelihood. It is the right to occupy and utilize land for a long period that can motivate people to manage natural resources with care. While security of tenure is generally identified as an important condition for sustainable natural resource management, there are divergent views as to the meaning of security in this context.

A common conception in governments and international agencies is that rights to natural resources must be properly registered, and that economic progress will only materialize under the conditions of private property rights. The underlying notion is that traditional tenure systems lack security and that the land is unsuitable for long-term investments. An opinion confirmed by anthropological observation is that traditional tenurial systems generally provide enough security for individuals and groups, while the farmer whose land rights are subject to modern law of tenure is often beset by insecurity (McCay and Acheson, 1987).

The important link between territory and kinship is often overlooked in discussions about land tenure. The individual is usually assured unambiguous rights to a particular parcel of land, through descent from a clan or lineage founder who secured the use-rights of his descendants. Indigenous land tenure is a bundle of rights, claims, privileges and liabilities associated with land. The same piece of land can be used for different purposes by different individuals and groups (for example cultivation, grazing, collection of fuelwood and wild plants, drawing water in springs or streams), and the regulations about use and access to common resources tend to be most elaborate in fragile environments. The rules and regulations are subject to interpretations, and they permit modifications and changes to comply with new situations. It is noteworthy that such changes are often inspired by environmental concerns and a rationale to restore sustainable environmental process.

The problem is that indigenous property rights are now in the process of being severely weakened and transformed through the integration of small communities into larger socio-economic systems at great opportunity cost (for example, Hjort af Ornäs and Salih, 1989). The shrinking resource base experienced by many individuals gives them no other option but to over-exploit the environment, usually against their own better judgement. A role for the rural sociologist is to identify traditional forms of land management, and to explore with local people ways in which they can be strengthened and modified to assist improvements in land husbandry. Part of the work is to listen and learn from local people about what the strategies of the indigenous inhabitants tell us (Richards, 1985), and what they can achieve by way of improvement with modest assistance, for sustainable natural resource management can only be realized by land users.

Gender and the environment

Gender is a characteristic that helps us to analyse the internal dynamics of a community. An understanding of the linkages between gender, poverty and the environment is necessary if we are to understand environmental processes and changes. There are direct linkages between women's and men's roles in social and economic processes on the one hand, and natural resource processes on the other. One example is land tenure where men in most cases hold those rights and where women obtain usufructuary rights to family and kin resources through marriage. This applies to family holdings as well as to the commons. The women's usufructuary rights to resources on the commons are the most vulnerable of all. These rights may be invisible to policy makers, implying a great risk that they may be eliminated in efforts to make land tenure regulations more formal and secure to individuals. If women lose usufructuary rights to, say, woodlands or springs, they lose access to the necessities of life. The effect of many people searching for alternative sources of supply from a diminished resource base is usually land degradation.

Protecting the environment entails a wide range of interventions, many of which presuppose a considerable labour input from the community. Conventional soil conservation structures are an example, for terraces have to be built and cut-off drains dug. Although some tasks are new, most of the operations required for soil and water conservation are allocated by gender in accord with tradition. Women perform most of the farm work, so we should expect them to be responsible for the bulk of conservation work. Great care must be taken so that environment is not added to the long list of women's caring roles, so reducing their ability to cope.

It is by understanding men's and women's interaction with the environment that we can identify non-exploitive solutions that combine environmental conservation with people's need to maintain production and diversify their livelihood base. Investigation to provide the information that leads to understanding is a role where anthropologists and sociologists can make a significant contribution.

Inter-disciplinary teamwork

The experiences of sociological and anthropological participation in practical development work provide a lesson for the future. It is more effective to have interdisciplinary teams including sociologists and anthropologists rather than having them provide an independent professional contribution. There are several benefits derived from a team approach. Anthropology and sociology become essential and influence core competence rather than being simply an add-on skill. Team participation also improves prospects for clear-cut and relevant recommendations for practical action about social issues.

As a team member an anthropologist or sociologist is able to sensitize and train colleagues on-the-job in application of sociological approaches and methods. Observations and analysis of socio-cultural and economic community characteristics, such as gender relations, opportunities and constraints for popular participation, local criteria for defining relative wealth and poverty and the general operation of the household or family economy, should be everybody's concern.

Integrating a Socio-economic Perspective into Soil and Water Management in Zambia

A.J. SUTHERLAND

Institutional background

Within the Research Branch of Zambia's Ministry of Agriculture and Water Development, the Adaptive Research Planning Team (ARPT) conducts farming systems research. Provincial teams for this comprise an agronomist, an economist, a research-extension liaison officer (RELO), and some provinces have a livestock specialist. Rural sociologists at regional level provide input to provincial activity.

A national team conducts soil research for the research branch. Extension in soil and water management is directed by the soil conservation specialists within the Irrigation and Land Husbandry Branch, and the land use planning officers at district level. The latter access small-scale farmers through area supervisors and camp officers of the Extension Branch.

Integration of sociologists

Social scientists (economists or sociologists) are only employed in the Research Branch, where rural sociology has been effectively integrated (GRZ, 1990), though positions for social scientists are not included on the staff establishment register. Social scientists occupying vacant positions dedicated to technical scientists feel marginalized by their status, and do not see prospects for career advancement. The absence of social scientists in the extension and Irrigation and Land Husbandry Branch means that socio-economic issues do not receive adequate attention there.

New challenges posed by soil and water management issues

Soil and water management research issues are a new challenge to ARPT. The farming system research methods used are well-suited to handling sustainability issues (Francis and Hildebrand, 1988), but there has been little work for the following reasons:

- Donor pressure to come up with quick and measurable results resulted in concentration on technology transfer thought likely to produce impact, for example new varieties and changes in cultivation practices
- The demand from extension branch for location-specific recommendations for field crops
- Farmers have not emphasized sustainability issues during problem identification activities
- The results from early attempts to address conservation issues by on-farm experiments were discouraging, for example benefits of soil liming in Central Province were not detected in experiments

- Research soil scientists have identified few technologies to test on farmers' fields, and
- Farmers' participation in the design and management of experiments has been limited.

Positive experiences

Some examples illustrate how a sociological perspective has been incorporated successfully at the provincial level.

Eastern Province
Eastern Province became the soil and water management focus for the extension branch from the mid-1980s because of observed land degradation symptoms, associated with the high rural population density. Sociological investigations were introduced recently, in response to poor adoption of soil conservation measures. In 1989, the Soil Conservation and Agroforestry Extension Project (SCAFE) initiated a survey into farmers' attitudes to soil erosion and deforestation (Sylwander and Egnell, 1989). A major finding was that, although women were likely to be involved in the implementation of conservation measures, the project was aimed at male farmers. Most farmers are aware of soil conservation problems and know of measures to tackle these, but this knowledge has been neglected. Social scientists were called in during an evaluation of the project and changes were introduced to involve women, and to train more farmers in the design and construction of physical conservation structures.

Sociological input from ARPT was made to an ethno-botanical survey conducted by an agroforestry research team in 1989. An outcome is that more local species are incorporated alongside exotic species in the research programme. A parallel on-farm study into the soil improving properties of these trees has been initiated.

Adaptive research has initiated a project to restore degraded lands. Social scientists have played a leading role in rural appraisal, including the training of commodity scientists and extension staff in informal survey methods (Sutherland and Singogo, this volume). This work was undertaken as part of the national research strategy to break down interdepartmental barriers, and to focus research on priority development problems (GRZ, 1990). A field survey identified a range of socio-economic factors influencing soil and water management. Household decision-making on crop choices, fallowing decisions, land tenure, marriage and inheritance systems, labour constraints, livestock ownership and access, access to seed, changing agricultural policies on pricing, input supply and credit, are all important because of their effects on land husbandry practices. More surveys are planned to explore the relationships between the sociological and technical issues. These will be conducted by interdisciplinary teams. Currently, social scientists are in the minority. The approach will be to increase involvement of technical scientists in identification and analysis of socio-economic information.

Luapula Province
The Luapula Province wetlands comprise a large but under-utilized potential resource for crop production (Kokwe, 1991). The ARPT team asked the rural sociologist to identify the reasons for the declining rice production. This study highlighted division of household labour, access to draught power, and agricultural credit policies, as important sociological constraints relating to the use of wetlands.

Lusaka Province
Drought was identified as the major constraint to agriculture and local food security during an informal survey conducted in 1984. However, the survey concentrated on rainfed crops and overlooked the widespread practice of using available and residual moisture through river bank cultivation during the dry season. After investigations ARPT sociologists showed the importance of wetland cultivation in the food system of the area. They also highlighted the importance of division of labour and land tenure in relation to the distribution of benefits from riverbank gardens.

Northern Province
Soil fertility has been the major thrust in on-station research activities for more than a decade. This research has had minimal impact on farming systems. Extension staff did not understand the soil classification system used by technicians, and this does not relate closely to the land use systems. In an attempt to bridge the gap between research, extension, and farmer knowledge, the ARPT rural sociologists have initiated a collaborative study with the soil survey unit into indigenous soil taxonomies. The objectives of this study are to compare indigenous and formal taxonomies, and produce a handbook to aid extension staff on soils related issues.

Lessons from the Zambian experience

Some lessons have been learned from the exploratory work in sociology. The lessons relate to socio-economic constraints and opportunities to improve land husbandry, and methods for conducting sociological research.

Socio-economic constraints and opportunities
It is recognized that appraisal of the community situation is the starting point for identification of opportunities for harnessing existing institutions and knowledge to bring about better land husbandry. The investigations require a participatory approach to technology development, and this presents challenges to the social scientist. It is important to identify some of the sociological factors which influence decisions. This will help those involved in participatory research and extension to identify potential issues arising during appraisal and evaluation activities. The following factors have been shown to influence adoption of land husbandry practices:

- Land tenure systems, particularly as influenced by the local systems of kinship and marriage and inheritance. Matrilineal systems generate a lot of geographical mobility, and the sharing out of property and separation

of male and female land rights pose special problems for technologies developed within a framework of title deeds
- Livestock ownership systems with skewed gender ownership, absentee ownership, and overlapping rights present complex issues relating to decision-making when approaching problems of overgrazing
- Food systems and cropping systems involve preferences about what to plant and what to eat. Systems dominated by starch staples, such as maize, which require high soil fertility, limit potential for crop rotation technologies
- Demographic forces, including population pressure, and imbalances between age classes and gender, impact on land tenure and cropping practices
- Local patterns of settlement and political organization impact on land use, and on planning and implementation of soil conservation measures. In Zambia there is a marked contrast between large nucleated villages with influential headmen, and scattered settlements with weak decision-making capacity
- Labour organization and practice merits careful study. Extension messages relating to tasks should be targeted at the right gender and age groups, and
- Agricultural policies impinge on practice. In Zambia, the removal of fertilizer subsidies has resulted in less fertilizer use. Credit provision has focussed on seasonal loans and there has been no incentive to adopt soil conservation measures which require investment and take several years to pay off. Moreover, loans have not contained conditions relating to improved land husbandry.

Approaches for sociological input
The following approaches have proved useful:
- Detailed in-depth research is the best way to investigate land tenure issues, but it is time-consuming and therefore costly
- Rapid informal surveys are useful for identifying constraints
- Formal methods can be used for collecting indigenous soil and tree classifications, and
- Aerial photo interpretation shows changes in settlement and land use over time.

Future prospects

The future prospects for incorporating sociological perspective into land husbandry issues depend to a considerable extent on the provision of adequate expertise. In the Research Branch the prospects are good because the role of sociologists has been accepted by management and scientists, and there is some experienced manpower available. In the Extension Branch there are no social scientists, and the approach remains technical. The importance of social issues is increasingly recognized and will lead to the eventual recruitment and establishment of manpower in both branches. There is a need for a training and recruitment strategy to foster the development of a national cadre of social scientists. This will require an investment in degree level training and the provision of appropriate career and incentive structures for social scientists working in the natural resource sector.

Economic Issues

Economic Considerations for Participatory Development of Natural Resources

J.P. HUNTER

The problem of non-sustainability

There are two central considerations in sustainable agricultural development and resource management. The first is popular participation in the identification of resource management problems and in the planning, implementation and evaluation of activities to overcome them. The second is to ensure technology is appropriate to the resource constraints of the local community and understandable to it. This approach recognizes that non-sustainable resource use is not primarily a technical problem that can be solved solely by technical fixes. It is a result of socio-economic factors that need to be addressed directly – a scale of activity that is amenable to local control.

However, as important as these criteria are they are necessary to sustainability but not sufficient. In addition, the activity must be economically viable. Economic viability determines the willingness of people to participate in sustainable development and shapes the choice of technology. It also refers to whether the society's scarce resources are being efficiently used, and whether the policy environment is supportive of resource sustainability. This must be evaluated at the household, project, and policy levels. Without economic viability, the most intense work at grass roots organizing, or the most clever technical solutions, will almost certainly come to naught.

Consistency at the rural household level

The immediate survival of many rural households in developing countries is dependent on their consumption of a mixture of material and cultural goods and services (food, clothing, housing, transportation, ceremonies, leisure, and so on). They may obtain these directly through home production (home-grown food, domestically-produced housing), indirectly through production for the market (agricultural produce, handicrafts), or by earning wages through sale of labour. Their future well-being is dependent on their ability to maintain and expand these options, principally through reproduction and investment (in land improvements, education).

Which options they take depend on culture and tradition, access to productive resources and the results of past investments in productive capacity. The decision is also influenced by the household's interaction with the economy at large, such as options for wage employment and wage rates, markets for agricultural goods and prices, credit opportunities and interest rates.

From another perspective, the rural household seeks to maximize its present and future well-being by allocating its resources to those productive activities with the highest payoffs. These resources include the household's access to farm and range lands, fisheries and forests, accumulated capital goods, and financial and labour resources. This latter is often the household's most binding constraint in southern Africa.

Households that have wage-earning options and face poorly-developed or highly unstable markets for agricultural output may be almost wholly subsistence producers. However, relatively few households in the modern world are pure subsistence producers. Most have some options for wage employment and market production, and some portion of the items contributing to the household's well-being is purchased. Some household production will be dictated by culture and tradition. Most, however, will have to compete for a household's time with alternative activities.

In much of southern Africa, wage-earning opportunities are many, and wages are high relative to prospects for farm income. Given the relative security of skilled male labour, farm operations will be selected which use unskilled, adolescent labour or female labour. Since female labour is heavily occupied in the production of domestic output and child care, it is available only sparingly for market production. Agricultural improvements will be adopted principally for their labour-saving, not for their yield-enhancing properties, since they may permit more labour to be freed for wage employment. Under these circumstances, agriculture is seen as a supplement to income earned elsewhere and as providing a livelihood after retirement.

Activities with a long-term payoff may not receive high priority. Long-term, labour-intensive conservation activities may be resisted. This is not because people fail to recognize their worth, but because that worth compares unfavourably with the cost of the labour that would have to be redirectd to these activities. Thus, activities which make sense in labour-surplus households in southern Asia, may not command themselves to relatively labour-short households in southern Africa. Efforts may need to be put to developing labour-saving, low maintenance conservation technologies. This could mean that innovative forms of labour mobilization and organization have to be employed.

Economies in which options for wage employment are few or poorly paid may offer greater scope for agricultural development and for conservation activities. Then, agriculture takes on a more central role in shaping the household's well being, and production for the market may be one of the best sources of a monetary income. When male household members are more likely to remain at home, more adult labour is available for agricultural intensification and for long payoff conservation activities. Assuming a relatively secure land tenure, the household may then be more willing to invest in conservation.

The above discussion emphasizes that improved resource management must be consistent with the constraints and motivations operating on rural households if it is to have a chance of success. Popular participation in resource management and techniques for involving farmers in decision-making and planning can help to identify and increase the awareness of these constraints and motivation.

Consistency at the project level

A key principle of economics is that an individual's, or society's, wants are essentially infinite while the means for satisfying those wants are limited. This is especially so in the less-developed countries, and post-structural adjustment Africa. To promote rapid development and avoid waste, it is incumbent that resources, natural or otherwise, be put to uses which maximize their net benefit to society.

The requirement that an activity's benefits must outweigh its costs (the principle behind cost-benefit analysis) is a key component of sustainability for several reasons. It helps to establish priorities and to put those priorities in perspective. Not all conservation goals are equally valuable nor, with the prevailing scarcity of resources, attainable. An evaluation of costs and benefits helps to identify those which are. Poor countries can ill afford to squander valuable resources on activities with little or no pay-off. In addition, a cost-benefit evaluation helps to distinguish appropriate from inappropriate priorities and means from ends. There is a tendency for some resource managers to think that resource preservation should be an end in itself. For resource users, however, the ultimate goal is the resource's contribution to human well-being. Cost-benefit analysis helps to keep this in the picture.

While cost-benefit analysis (CBA) is in principle non-controversial, in practice it must be done with care to avoid pitfalls. Aside from the fact that it may be done poorly or used deceitfully to justify already-determined intentions, there are a number of components of CBA about which there is considerable controversy. This is especially true of resource and environmental management for the following reasons.

Externalities

Key features of environmental problems are their widespread impacts and spillover effects. The results of environmental degradation frequently impose costs on people far removed from the site of the degradation and who have little or no direct responsibility for its cause. Conversely, efforts to improve environmental management may provide benefits to people who took no part in their improvements. These spillover effects are called externalities because they are external to the persons most directly responsible for resource management.

Externalities are important to the sustainability of resource management efforts for several reasons. Negative externalities, in which users do not fully bear the costs (including long-term degradation costs) of resource use, are the principle reason for non-sustainable, excessive resource use. The presence of negative externalities are a distinguishing feature of common property resources such as rangelands, fisheries, forests, and are a major factor in pollution. Positive externalities, in which the benefits of resource improvements are not fully captured by the improver, are a major cause of inadequate resource improvement. Both types of externalities can be corrected through the careful definition and enforcement of property rights, and through the creation of markets for the resource in question.

Extended pay-off
Another key feature of natural resource investment is that it often pays off only over a long period of time. As a result, its benefits may not be readily apparent, or viewed as not important enough to compel resource users to alter their behaviour. This problem may be especially severe if users are resource-poor and value short-term gain over long-term benefit. Despite this, policies to lower interest rates and choice of appropriate discount rates for project evaluation may overcome much of the problem.

Non-quantifiable benefits and costs
Many of the pay-offs of environmental improvement are non-quantifiable and difficult to measure. With some (such as, quality of life, cultural preservation, a pristine environment), the problem is inherent in the nature of the benefit and only qualitative, not quantitative, values can be attached, although economists are becoming increasingly innovative in determining values for benefits that had been previously thought non-calculable.

Distributional issues
The valuation of an activity may depend on the distribution of costs and benefits among different groups in society. Thus, activities with only long-term benefits may not be highly valued by the poor. Easier credit or targeted benefits may be necessary to overcome this.

These problems should not be taken as arguments against such analysis. If quantifiable benefits and costs are weighed against each other explicitly, more rational argument can then take place about the relative merit of non-quantifiable benefits and costs. Cost-benefit analysis forces one to set priorities and identify benefits and costs explicitly, and it makes the justification for an activity more rigorous. While cost-benefit analysis is not a guarantee that society's scarce resources will be used efficiently, if it is done carefully, openly, and by trained practitioners, it can minimize the danger that they will be used inefficiently.

Consistency at the level of the economy
Policy choices made at the macro-economic level can affect the viability of activities at the farm and project levels. Often these policies have no ostensible relation to resource use. Nonetheless, policies which affect output or input prices, interest rates or credit policy, wage rates, exchange rates, imports and exports and rural-urban differentials in living standards can all affect the types of resources used, their use rates and, ultimately, the sustainability of that use.

Markets are the primary transmission channels of economic information (prices, wages) to economic agents. With this information, decisions can be made on what and how much to produce, and by what method, and when to consume. Changes in this information may stimulate new consumption choices, production processes and technologies. Market information may also transmit the need for regulation to public officials. If they work properly, markets are potentially the most efficient means by which this information can be conveyed.

Natural resources pose special problems for markets. The first problem is one of existence. Markets may not exist because of the nature of the resource and the difficulty or expense of establishing definable ownership or tenurial rights to it. As a result, the resource is held commonly, and users have open access to it. In some cases markets fail to exist because traditional rights of communal access do not permit them. Situations in which it is difficult or undesirable to exclude potential free-riders make it difficult for markets to exist.

Another problem may be market imperfections, of which oligopoly and monopoly are important. Administrative prices set by marketing boards or parastatals, or subsidized prices that reflect political, not economic, realities may create distorted markets. Although markets exist and price signals are transmitted, the signals do not accurately convey the true social scarcity value of the resource but a distorted, artificial value.

Where externalities exist, markets will not adequately account for all of the benefits of resource rehabilitation or environmental improvement nor all of the costs of resource misuse. Free-riders are plentiful, and there will be too little improvement and too much misuse. Sometimes some institutional ingenuity can devise ways of internalizing these externalities.

Uncertainty about the future, combined with an unwillingness to take risks, may result in short-sighted decisions about resource use. This may be particularly damaging where environmental destruction is irreversible, or resource users are so poor that they must put short-term survival ahead of long-term environmental concerns.

Some, particularly those philosophically opposed to markets for natural resources or environmental attributes, have seen these problems as justification for abandoning market incentives entirely when trying to solve environmental problems. Instead, they have opted for administrative sanctions, prohibitions, and fines. Market imperfections should be viewed as necessary, but not sufficient, conditions for government intervention. Intervention is only justified if it can improve the market or is less costly than the market failure it is designed to correct.

Regulations and fines are likely to be more expensive in terms of administrative and enforcement cost and political ill-will than self-interested compliance. They are also likely to be less effective and efficient. Even where markets for the resource itself are difficult to create or deemed undesirable, it may still be possible to create market valuation of environmental attributes. Government can establish limited, negotiable rights (for example, grazing rights) to these attributes and set up markets whereby these rights can be traded. Thereby, resource users are forced to bear the costs of their use.

In addition to market failure, policy failures may distort incentives in favour of over-exploitation and against conservation and so exacerbate resource misuse. These policy failures include taxes, subsidies, and quotas and are often widespread and pervasive. Many areas of policy failure may be far removed from conservation or resource management activities, such as exchange rate distortions, excessive interest rates, and high inflation. Trade distortions, especially those that artificially protect industry may worsen the agricultural-industrial terms of trade, and make sustainable agricultural production relatively unprofitable. Structural adjustment programmes, which promote economic growth through increased trade performance, can have an adverse impact on resource management and environmental quality if policies which neglect to account for the costs of resource degradation in their cost-benefit analysis are adopted. Similarly, care must be exercised in determining which crops to emphasize and which technologies to adopt when embarking on agricultural adjustment programmes.

Conclusion

Many past attempts to correct environmental degradation and promote conservation activities have failed to recognize that there are usually economic reasons for prevailing patterns of resource misuse. Resource users have been accused of being ignorant about the imperatives of proper resource management and their own long-term interests. Sometimes, they have been accused of wilful misuse. Coercion, bans, and sanctions are instituted in an attempt to change behaviour. Although this was the pattern of colonial times, it has continued all too often in the post-colonial era. The result is that government is saddled with unpopular regulations difficult to enforce, resource users become embittered and rightly feel that regulators do not understand their needs. Thus, non-sustainable resource use continues.

Recently, efforts to correct this colonial legacy have focused on issues of sustainability and people's participation. This has provided a useful corrective to the earlier authoritarian, coercive, and elitist control measures. However, unless the underlying economic rationale for existing practices is recognized and corrected, this new focus is likely to be as unproductive as the previous one. Corrective policy measures need to be taken at the national level. In particular the impact of negative and positive price incentives must be recognized and these must be harmonized with the requirements of sustainability.

Next, projects and policies must be honestly subjected to cost-benefit analysis. Poor countries cannot afford to squander resources on activities with poor payoffs. Finally, activities must recognize the special motivations and constraints of resource users. Corrective measures must be consistent with these if they are to have hope for success.

Economic Management of Natural Resources by Rural Communities

N. REYNOLDS

Introduction

Southern Africa is emerging from a quarter century of political, social and economic turmoil, and majority demands for political independence, social equality, and economic development are finally being addressed. What prospects might these changes have for the sustainable use of natural resources and the creation of environmentally responsible societies?

The history of much of post-independence Africa has been traumatic. The failure of African governments to guarantee democratic political and economic rights has resulted in political instability, poor economic performance, and the pursuit of short-term political and economic objectives at the expense of long-term sustainability. Widespread environmental degradation has been a result.

In an attempt to reverse the poor economic performance, governments are embarking on economic reform programmes that reduce subsidies and cut the levels of real expenditure on social services. Rural institutions and economies will have to shoulder a heavier burden in the provision of social services and in resource management if past gains are to be preserved and extended.

Reform of the countryside

A key feature of the colonial history of most southern African countries was the alienation of the best agricultural land from the indigenous population to immigrant settlers. This was accompanied by a similar alienation of rural communities from the ownership and control over their indigenous wildlife and forest resources. Officials thought them incapable of properly managing these resources and, in an effort to preserve and protect them, departments of wildlife, national parks, and forestry progressively restricted local access and control and placed resource ownership in the hands of the state. Limited residual rights remained but could be withdrawn at the will of the state. As outsiders were granted licenses to exploit these resources, locals increasingly became poachers on state land, the enemy of wildlife, and a threat to forests. Professional resource managers became policemen. Antagonism between government field agencies and local communities was the rule, and increased rural poverty the result.

This alienation of communities from some of their key resources was paralleled by the shift from land abundance to land shortage. Colonial regimes transformed villages from active, self-governing entities to passive, administered ones. Village institutions were co-opted for the purposes of law and order and, in the process, became static and inflexible. Just when they were needed to provide adaptations to the changing resource balance, their ability to do so was restricted.

With independence there was a chance to redress some of these probems of land and resource alienation. In most countries the administrative focus on law and order was maintained, but turned to support new political needs for information and control. Although reforms to local government systems often featured the replacement of traditional authorities by elected village committees there was rarely any attempt to address the altered man-land position. In the rural areas, power remained hidden in informal institutions. Rather than being helped to take charge of their resources, villages were encouraged to look to central government for their needs.

In each country there is a pressing agenda in the countryside. It is to raise production and productivity, to preserve and extend the delivery of social services, to halt and reverse ecological degradation, and to generally improve the capacity of communities to manage their own resources. Rural economic security must be restored so that villagers can move from short-term survival to sustainable long-term production. This will require revitalized rural institutions that alter the man-land relationship from one of exploitation to one of husbandry, and an effective redistribution of power and responsibility to the countryside. If this agenda is met successfully, an increasingly competent rural citizenry will be better able to assist governments in securing viable social and economic services.

The change from passive, administered communities to active and cohesive managers of land, water, forests, wildlife, and the tourist and trophy hunting businesses, is not easy. The decision to decentralize control from state to community is often grudgingly given. In all governments old attitudes persist. Paternalism, a feature both of colonial and of post-independence centrist ideologies, reinforces a reluctance to relinquish central control of revenues. Structural adjustment programmes with their emphasis on increased revenues, exacerbate this. Nonetheless, change is necessary if the problems of the countryside are to be successfully addressed.

An emerging solution

There has been a growing realization during the 1980s that the surest way to conserve nature is to establish clear rights over land and other natural resources under modern law. In Africa, where communal rights to resources are the norm, these rights should be communal and build upon traditional rights. It is particularly important that the right of exclusion be firmly established since many African land management systems properly leave land idle for varying periods and, without exclusion, there may be a tendency for the seemingly idle land to be encroached.

Centred upon Africa's unique resources, wildlife and tourism have provided a stimulus for change and reform which has encouraged local people's involvement in the effective management and preservation of their wildlife resources. This has heightened old land use conflicts, brought the perennial tension between local rights and central revenues to the fore, and created a demand for local control over newly-valuable wildlife resources.

In some countries, communities have been drawn into conservation through the appointment of village game guards. The guards receive a stipend, usually operate under the local chiefs, and report to the national parks service. This service works well against outside poachers. The guards benefit from the observation and reporting of the community about strangers and their movements. Sometimes a share of tourist and hunting receipts is made payable to communities bordering national parks. As yet, few communities participate in schemes that generate revenue from tourist or safari operations on their land.

Recently in Zimbabwe, the authority for the management and utilization of wildlife has begun to be passed by government to communities, which then enjoy rights similar to private farmers. They can decide how they will operate wildlife and tourist businesses, alongside cattle and cropping. Usually they enter into contracts with safari or other operators since they themselves do not have the experience to sell safaris abroad and to deal with special clients. The results, still preliminary, are encouraging. Until given responsibility, no one can be expected to act responsibly. To complete the logic, communities must be the direct beneficiaries of their husbandry of resources. In Zimbabwe and Namibia newly organized communities are pooling land together with commercial farmers and national park administrations to form large conservancies for wildlife and tourism businesses.

The democratic company model

The Rural Structure Adjustment Programme
A broad consensus is emerging that villages are capable of managing a wide variety of resources and activities, including tourism. In Zimbabwe the Association of District Councils has adopted a reform programme centred upon the transformation of the village into a democratic company. Termed the Rural Structure Adjustment Programme (RSAP), to complement the national economic reform programme, it alters the definition of village membership from a birthright that guarantees access to the land, to membership in the ownership of the village assets. The group interest becomes husbandry of, and investment in, jointly-owned assets. The democratic company is essentially a property company.

Equality for women
A significant milestone is the inclusion of women as village members, giving them equal ownership shares with men. This vital reform, which recognizes the important roles of women in farming, is necessary because the household as a membership unit is unsatisfactory, for there are inequalities brought about by differences in size, capacity, and changes over time.

Allocation of rights to resources

The democratic company model is an evolution from the traditional right of equal access to use and exploit the land, to equal rights of membership expressed through common rights to the resources of the village. Each year, members are allocated equal rights to the use of the common village assets, such as arable land, grazing land, water, and woodlands. These rights may then be traded among themselves according to each individual's resource needs. In the process, resources acquire values which reflect the balance between their relative abundance and the demands placed upon them. Those who use less are compensated by those who use more.

The process can be illustrated with grazing rights. Previously, outside experts claimed villagers were ignorant since they overgrazed and destroyed their land. The outsiders never provided a model as to how a village could limit the number of livestock carried, and distribute that number among members in an equitable and efficient manner. In the company model, the community can decide what number of animals to carry (for example, 1000 livestock units). Annually the number of animals is divided by the number of members (for example, 100) to provide equitable shares. The members trade the shares amongst themselves, realizing a price for grazing (such as $10 per unit). The value of the grazing is $10 000 to the company and $100 to each member. At the annual general meeting it may be decided to retain a portion of the total tradeable value for reinvestment. Each member has to contribute in proportion to his or her share holding. Poor members sell grazing rights to realize the monies due.

The pricing of resources gives the community an annual measure of village productivity, and experience of financial and resource management. The majority of members, being small producers, will favour high prices, asset conservation and husbandry. While the few larger farmers might seek lower resource costs. As a fully democratic company, the majority view should prevail, but with a concern for efficiency so that producers are rewarded.

The model lends itself to the management of all common resources. It provides members and the community with management tools in the form of land use prices, and with the means to optimize individual and communal decisions. Having agreed the value of each land use, communities can talk to banks about the possibility of loans to fund commercial investments, since they have the means and mechanisms to assure repayments.

Assets in trust

In Zimbabwe under the RSAP, all assets are vested in a village trust. Only the cash flow is available to be pledged against bank loans. Hence only a fraction, perhaps thirty per cent, of present cash flow can be safely pledged for repayment. Government and donors have been asked to support the RSAP by providing matching grants equal to three years of retained profits.

Taxation and representation of communities
Villages will submit to a local district council tax of twenty per cent of the profit distributed to members. Villages have need of improved and maintained infrastructure, and only the council can provide the regional assets and services.

Under the RSAP, reformed village companies are serviced at the district, and represented at the national level by their own village development society. The society will help to distinguish between the line functions of council and of ministry departments, and the economic, resource management, and production functions of the village. The intention is to force a constructive reappraisal of the relationships of villages and government.

Conclusion

The work to restore community control over wildlife has reached a stage of take-off, and for some communities considerable benefits may accrue in coming years. Village institution building exists more in the form of slogans than in practice, but centrist political and administrative concerns are being slowly replaced by imaginative search for working answers. It is that process that needs reinforcement to enable an increase in local control and management of natural resources.

The RSAP programme is showing real potential, but this success brings the danger of rapid replication ahead of the capacity to service and support communities that vote to adopt new responsibilities for resource management. Official support for the launch of the RSAP is being sought for all districts in Zimbabwe. A new national, and even regional, ethos is being generated which seeks to alter the political and administrative environment in which communities can explore new expressions of rights, and new forms of resource management and of representation.

Participatory Appraisal, Planning, and Development

Introduction

R.J. CHEATLE

Four papers in this section present a perspective on participatory rural appraisal (PRA) and rapid rural appraisal (RRA). Many practitioners agree that a diagnostic appraisal of some kind is a prerequisite to development activity. The purpose is to obtain a precise statement of on-farm needs, so that scarce resources can be applied to secure development impact. The work of appraisal includes identification of on-farm problems, agreeing priorities, and the setting of strategy and tactics. These design aspects should be accomplished before research and extension activities are commenced, to ensure that support services address significant problems.

Rapid rural appraisal (RRA) is a well-established form of diagnostic appraisal with multi-disciplinary implementation and participation as two essential features. Targeted fieldwork conducted by multi-disciplinary teams provides an integrated and effective approach to collection and analysis of data, and project design, are the main uses of RRA to date. The advanced forms of RRA go well beyond interviews for extracting data. They seek to learn from and with farmers, and are therefore participatory, because people are encouraged to articulate their problems. Yet, most decisions about any post-appraisal activity usually remain as the prerogative of the development professionals.

The evolution of PRA out of RRA signals an important shift towards local decision-making, control, management and responsibility for natural resources. This is because, in PRA, farmers are encouraged to become more responsible for their own analysis, project design, experimentation and evaluation. The professional role in this development partnership changes to facilitator and provider of development opportunities. The intention is that professionals provide expertise, but that farmers take decisions about what to do. An important goal initiated by PRA is the identification and implementation of sustainable natural resource management plans by the target group themselves. This change in approach offers exciting potential.

For example, all three supporting authors to this section illustrate ways in which farmers integrate conservation practices into their production systems without any assistance from support services. Accepting the principle of local responsibility for natural resource management by land users, and using professionals to suggest how improved practices can be built on to what is already being done, will inevitably lead to improved land husbandry. Any change agreed by farmers will bring benefits in its wake. Farmers are smart enough to adopt sustainable practices that bring advantage.

Obtaining farmer adoption of recommended production practices is a major stumbling block in development, for much of what is researched and recommended is ignored. PRA is holistic, and encourages socio-economic consideration in respect to both production and conservation. By encouraging the articulation of real feelings PRA aids understanding of the socio-economic rationale for what people do. With a degree of understanding it becomes possible for professionals to help match solutions to farmers' problems, and for everyone to agree practical changes that make sense. If a new or revised practice increases benefit the farmers are likely to adopt it.

There will never be enough researchers and extensionists to develop and transfer messages to service all farmers. A successful transformation of roles initiated through PRA could allow farmers to take on some responsibilities for research and extension, and to be helped in this work through partnerships with development professionals. Implicit in this concept is a potential for unleashing enormous power to influence development positively, by emphasizing professional roles as facilitators, by mobilizing more people to research practical improvements on-farm, and by bringing to public notice good practices through improving the mechanisms for farmer-to-farmer information exchange.

Promoting organized and managed activity through the allocation of functional roles is basic to economic success. By placing emphasis upon communities and interest groups to create and improve organization and decision-making, PRA switches attention from the individual farmer and helps provide a foundation for a general commitment to sustainable action and economic improvement.

The thrust of PRA is for decentralization, and the provision of mechanisms that enable people to command and manage their resources. Helping to build confidence, organization and capability to consider and deal with conservation issues and resource management at the grassroots level will help improve practical land husbandry.

PRA is at a very early stage of development. Operational methods are still being invented and refined, and the potential of groups and individuals to contribute to development is being assessed. The results so far indicate considerable potential and scope. PRA has been used effectively by expert facilitators to assist in project design. Development of methods for participatory monitoring and evaluation (PM and PE) is a field that needs attention, so that we can learn about the reality of increased client independence in development. These essential processes should incorporate in-depth consideration of the technology adoption and dissemination processes. Through high profile case studies, implementation of PRA can lead us to a better understanding of the processes of change in the rural environment and optimum roles for facilitators and land users. All this is vital knowledge if good land husbandry practices, worked up and adopted as a consequence of PRA by a few target farmers, are to be taken up by many farmers.

A paper contributed from Zambia is a brave example of people taking a step towards participatory development by gaining experience in diagnostic survey. Typical problems faced by those intent to investigate the potential of rural appraisal are illustrated. The problems are burdensome, partly because organizational structures are not matched to a set of functions required for either multi-disciplinary or inter-disciplinary activity in concert with farmers, and also because money and experience are scarce resources. Ideally, a review of functions would precede changes, so that structures can be established to meet functional needs. It is unlikely that major agencies will take radical steps of review, and change is likely to be piecemeal as confidence in appraisal and participatory methods grows.

A good deal of work is necessary to develop methods of participatory development and to demonstrate a significant correlation between application of methods and indicators of developmental success. However, when conducted by experienced personnel, RRA and PRA are already shown to be useful for diagnosing on-farm needs and development directions. The most immediate need is to provide training and hands-on experience for many more people, so that appraisal can be more widely used.

Participatory Rural Appraisal

ROBERT CHAMBERS

Evolution of rapid rural appraisal

The philosophy, approaches and methods now known as rapid rural appraisal (RRA) began to coalesce in the late 1970s. There was growing awareness of the biases of rural development tourism, and of the costs, inaccuracies and delays of larger-scale questionnaire surveys. More cost-effective methods were sought for outsiders to learn about rural life and conditions.

In establishing the principles and methods of RRA, people and institutions from many countries contributed. Numerous sub-groups produced ideas, tested them and reported their results. Among these, Gordon Conway and others (Gypmantasiri *et al.*, 1980) pioneered agro-ecosystem analysis (Conway, 1985) at the University of Chiang Mai in Thailand. In the mid-1980s the University of Khon Kaen, also in Thailand, was a world centre in developing theory and methods, especially for multi-disciplinary teams and for institutionalizing RRA as part of professional training (Khon Kaen, 1987). In the late 1980s, the International Institute for Environment and Development in London became a leading voice (McCracken *et al.*, 1988). In health and nutrition a parallel movement, drawing on social anthropology, was evolved in the 1980s under the rubric of rapid assessment procedures (RAP) (Scrimshaw and Hurtado, 1987).

Towards the end of the 1980s, numerous RRA approaches and methods were eliciting a range and quality of information and insights inaccessible with more traditional methods. RRA was argued to be cost effective, was more generally accepted, and gave rise to participatory rural appraisal (PRA) (Mascarenhas *et al.*, 1991) which evolved, and continues to evolve, rapidly.

Principles and practice of RRA and PRA

RRA is mainly extractive. Outsider professionals go to rural areas, obtain information, and then bring it away to process and analyse. PRA, in contrast, is participatory. Outsider professionals still go to rural areas, but their role is more to facilitate the collection, presentation and analysis of information by rural people themselves. With RRA the data is owned by the outsiders, and often not shared with rural people. With PRA data is owned by rural people, but usually shared with outsiders.

That said, good RRA and PRA have features in common. Most practitioners would agree on the following:

- *A reversal of learning* – outsiders learning from and with rural people, on site, and face to face. Rural people's criteria, categories, and priorities, and their indigenous technical knowledge are elicited
- *Learning is rapid and progressive* – conscious choice and flexible use of methods to explore important questions as they arise, with improvisation, iteration and probing
- *Trade-offs* – sought between quantity, accuracy, timeliness and relevance of information
- *Triangulation* – used to crosscheck and confirm data and to improve approximations, using several, often three, methods of sources and information
- *Optimal ignorance is sought* – meaning not trying to find out more than is needed, and not making inappropriately precise measurements. The collection is avoided of data that will not be used
- *Biases are recognized and offset* – for example biases of movement and contact which are spatial (where outsiders go), institutional (what organizations they visit), personal (who is met) and temporal (when they go, by seasons and time of day). Special efforts are made to meet those, often women and the poorer, and who tend otherwise to be missed, and
- *Team composition balanced* – in terms of gender, discipline, and other dimensions, and team interactions are consciously managed.

Beyond these common factors, PRA has added others which have not been prominent in RRA. These include:

- *They do it* – facilitating investigation, analysis, presentation and learning by rural people themselves, so that they own the outcomes. This often entails starting a process and then sitting back and not interviewing or interrupting
- *Self-critical awareness* – meaning that practitioners are continuously examining their behaviour, and trying to do better
- *Relaxing and not rushing* – exploiting the paradox that taking plenty of time in PRA is often faster and better than trying to be quick
- *Embracing error* – meaning welcoming error as an opportunity to learn to do better
- *Using one's own best judgement at all times* – meaning accepting personal responsibility rather than vesting it in a manual or a rigid set of rules, and
- *Sharing of information and ideas* – between rural people, between them

and practitioners, and between different practitioners, and sharing camps, training and experiences between different organizations.

The menu of methods for RRA and PRA

In its early days, RRA seemed little more than organized common sense, and some of the methods had been in use already. During the 1980s, though, much creative ingenuity was applied and more methods invented. In approach, there is a distinction between RRA and PRA; but most of the methods are shared. An indication of the range and variety of RRA/PRA methods can be given in a listing of some of the more common ones:

- Secondary sources – such as files, reports, maps, articles and books
- Do-it-yourself – asking to be taught to perform village tasks, such as transplanting, weeding, ploughing, field-levelling, drawing water, washing clothes, thatching
- Key informants – asking 'who are the experts?' and seeking them out
- Semi-structured interviews – this has been regarded by some as the core of good RRA. It can entail having a mental or written checklist, but being open-ended and following up on the unexpected
- Groups of various kinds – (casual, specialist, deliberately structured, community/neighbourhood). Group interviews and activities are often powerful and permit crosschecking of information
- Sequences or chains of interviews – from group to group, or group to key informant, or a series of key informants, each expert on a different stage of a process (for example, men on ploughing, women on transplanting and weeding, and so on)
- Villagers and village residents as investigators and researchers – women, school teachers, volunteers, students, farmers, village specialists, poor people. They can do transects, observe, and interview other villagers
- Participatory mapping and modelling – in which rural people make social, demographic, health, natural resource (soils, trees and forests, water resources and the like) or farm maps, or construct three-dimensional models of their land
- Participatory analysis of aerial photographs (often best at 1:5000)
- Transects – systematically walking with informants through an area, observing, asking, listening, discussing, identifying different zones, local technologies, introduced technologies, seeking problems, solutions and opportunities, and mapping and diagramming resources and findings
- Time lines – listing major remembered events in a village with approximate dates
- Local histories and trend analysis – people's accounts of the past, of how things close to them have changed, ecological histories, changes in land use and cropping patterns, changes in customs and practices, changes and trends in population, migration, fuels used, education, health, and so on, and causes of these
- Seasonal diagramming – days and distribution of rain, amount of rain or soil moisture, crops, agricultural labour, non-agricultural labour, diet,

food consumption, sickness, prices, animal fodder, fuel, migration, income, expenditure, debt
- Livelihood analysis – stability, crises and coping, relative income, expenditure, credit and debt, multiple activities, and so on
- Participatory diagramming – of flows, causality, quantities, trends, rankings, scorings, in which people make their own systems diagrams, bar diagrams, pie charts, and estimates using seeds, pellets, fruits or stones as counters, sometimes combined with participatory maps and models. *Chapati* or Venn diagramming is a method for identifying individuals and institutions important in and for a community, and their relationships
- Well-being or wealth ranking – identifying clusters of households according to well-being or wealth, including those considered poorest or worst off
- Analysis of difference – especially by gender, social group, wealth/poverty, occupation and age. Identifying differences between groups, including their problems and preferences
- Contrast comparisons – asking one group why another is different or does something different, and vice versa
- Ranking and scoring, especially using matrices and seeds to compare through scoring, for example different trees, or soils, or methods of soil and water conservation, or varieties of a crop
- Key local indicators – such as, what are poor people's criteria of well-being, and how do they differ from those we assume for them
- Key probes – questions which can lead directly to key issues such as 'What do you talk about when you are together?' 'What new practices have you or others in this village experimented with in recent years?' 'What vegetable, tree, crop, crop variety, type of animal, tool, equipment . . . would you like to try out?' and 'what do you do when someone's hut or house burns down?'
- Case studies and stories – a household history and profile, coping with a crisis, how a conflict was or was not resolved, and the like
- Team interactions – changing pairs, evening discussions, mutual help, and so on, where the team may be just outsiders, or a joint team with villagers
- Presentations and analysis – where maps, models, diagrams, and findings are presented by villagers, or by outsiders
- Brainstorming – by villagers alone, by villagers and outsiders together, or by outsiders alone, and
- Report writing at once – either in the field before returning to office or headquarters, or by one or more people who are designated in advance to do this immediately on completion of the RRA or PRA.

Practical tips

It is easier to give advice than to take it. Here is a personal list of some practical tips:
- Don't lecture. Look, listen and learn. Facilitate. Don't dominate. Don't interrupt or interfere. Once a task has been initiated let people get on with it

- In a discussion give people time to think or discuss among themselves
- Embrace error. We all make mistakes, and do things badly sometimes. Never mind. Don't hide it. Share it
- Try to obtain opinions from all groups, especially those – women, the poorer, the more remote, those who do not use services – who are liable to be left out
- Observe keenly to determine whether your eyes confirm the information given
- Relax, don't rush
- Meet people when it suits them, and when they can be at ease, not when it suits you, and don't force discussions to go on for too long. Stop before people are tired
- Be around in the evening, at night and in the early morning. Stay the night in villages if you can
- Allow unplanned time, walk and wander around
- Ask about what you see
- Probe. Often we accept the first reply to a question as being all that is needed, when there is much to be learnt, and people know more than we suppose
- Use the six helpers – who, what, where, when, why and how
- Ask open-ended questions
- Show interest and enthusiasm in learning from people
- Have second and third meetings and interviews with the same people and discuss your findings
- Allow more time than expected for team interaction, and for changing the agenda, and
- Enjoy it! RRA is interesting, and should be fun for all.

More about methods and other aspects of RRA and PRA is to be found in the further reading references at the end of this book.

Participatory rural appraisal

PRA is facilitated by outsiders but more than RRA it involves rural people themselves in investigation, in the diagramming, presentation, analysis and ownership of information, in the identification of preferences and priorities, and in planning, action, monitoring and evaluation. Thus, it is a new form of RRA which shifts the initiative and the action from outsider to insider, from the trained professional to the rural people themselves.

Besides agro-ecosystem analysis, with its special strengths in observation and diagramming, and earlier RRA with its special strengths in semi-structured interviewing and multi-disciplinary team management, PRA has other antecedents. These include activist research, in which under-privileged people are encouraged and enabled to analyse their conditions. Applied social anthropology, with its participant observation and its distinction of *emic* (insider) from *etic* (outsider) values, categories, and views of reality, and farming systems research, with its understanding of the complexity and diversity of farming systems are both important here.

The term PRA was probably first used during 1988 to describe village-level investigations, analysis and planning undertaken in Kenya (NES *et al.*, 1990). A parallel movement in India was taking place, notably with the Aga Khan Rural Support Programme in Gujarat (McCracken, 1988), and subsequently more and more NGOs in India have evolved and applied PRA, with MYRADA, Bangalore playing a prominent part in development, dissemination, and training government organizations. Much of the Indian experience was captured in *RRA Notes Number 13* (Mascarenhas *et al.*, 1991) which reported on a PRA workshop convened by MYRADA in Bangalore in February 1991. In 1991 and early 1992 there was an increasing interest in many countries. Among other forms of south to south sharing, trainers and practitioners from Kenya, Nigeria, Senegal, Sudan, Tanzania, Zimbabwe and other countries visited India, hosted by Action-Aid, AKRSP and MYRADA, for familiarization with approaches and methods current in India.

The core of PRA

PRA takes different forms in different countries and organizations. In Kenya, one form is described in a manual, and is linked to the production of a Village Resource Management Plan (NES *et al.*, undated). Another form has been developed by the soil and water conservation branch of the Ministry of Agriculture in conjunction with the International Institute for Environment and Development (Pretty, 1990). In India, a plurality of forms has evolved, but with a common underlying philosophy and experience. Drawing on various Indian and IIED approaches, a core of PRA can be suggested. In this, villagers' capabilities, outsiders' rapport, and visual sharing are key features.

Villagers' capabilities

Good PRA empowers villagers by encouraging them to take the lead in investigation and analysis. Rural people have shown a greater capacity than outsiders have expected to map, model, quantify and estimate, score, diagram, and analyse. When local materials are used, like the ground for mapping and diagramming, or seeds for quantifying, estimating and scoring, participation is often uninhibited and relaxed, with a willingness to express, share, crosscheck, and analyse knowledge.

Maps made on the ground or on paper have been of several types: social maps, indicating households and social groups; demographic and health maps, indicating women, men, pregnant women, children who are breast-feeding, immunization staus, handicapped, drunken husbands, and so on; well-being or wealth maps, showing who is considered well off, badly off and so on; natural resources maps, including soils, land types, forests and trees, and water. Matrices can be used for scoring and ranking items, such as types of trees, varieties of a crop, methods for soil and water conservation, or the characteristics of local categories of soils. Diagrams have been used by rural people to present and analyse many types of information, including nutrient flows on a farm, the impacts of a project, important people and institutions in a community and outside it, and the services used by villagers. With estimating and quantification also, as with maps, models, matrices and diagrams, villagers have shown themselves capable of presenting and analysing information far beyond normal professional expectations.

In all these, the methods and materials used have helped villagers to express and analyse their knowledge, but methods in themselves are not enough.

Rapport

The key to facilitating such participation is good rapport. For this, the behaviour and attitudes of the outsider as facilitator and catalyst are basic. Some of the keys are listening and learning, taking a keen interest, joining in activities, and patience. Fieldworkers now ask villagers to teach them, show respect and encourage people to have confidence in their ability. The fieldworker becomes a facilitator and hands over the lecturer's stick to become a student.

Visual sharing

Visual sharing is a common element in PRA. With a questionnaire survey, information is transferred from the words of the person interviewed to the paper of the questionnaire schedule where it becomes a possession of the interviewer. In contrast, with visual sharing of a map, model, or diagram, all who are present can see, point to, discuss, manipulate and alter physical objects or representations. In participatory mapping and modelling, villagers draw and model their villages and resources, deciding what to include, and debating, checking and modifying detail. Everyone can see what is being said because it is being done. The ground and local materials used have the advantage of being theirs, media which villagers can command and alter with confidence. Those who take part themselves learn from what is shown. Maps, matrices and diagrams provide visible checklists and agendas, and can be interviewed, providing a basis and focus for reflection and planning.

Training and problems to be addressed

A current question is what should be done about training. Such is the interest in PRA that demands for training of government and NGO staff are increasing. Three important problems are presented by this perceived need.

Training courses tend to formalize, codify and standardize, often in the name of quality. PRA requires flexibility and rapid adjustment to seize opportunities in impromptu situations. The lack of a manual for PRA in India has been much of its strength, for would-be practitioners have been forced to learn, not from books but through sharing and from their own field experiences. Many of the best innovations have come from practitioners not following established methods. It is not books of instruction but personal commitment, critical awareness and informed improvisation and creativity, which assure quality.

Faddism is another problem. Like farming systems research, RRA and PRA could be discredited by becoming too fashionable too quickly, and then adopted too fast and misused. The warning signs are there: demand for training which far exceeds the competent trainers available, requirements that consultants use PRA and then, consultants who say they will do so, but do not know what PRA entails, and the belief that good RRA and PRA are simple and easy, quick fixes, which they are not.

The word rapid is now a liability. It is in danger of being used to legitimize biased rural development tourism. It would be better if the first R of RRA stood for relaxed, implying plenty of time. Hurry usually means that the poorest are not met, listened to, nor learnt from, when much of the rationale for RRA/PRA is to learn from them, and to empower them.

RRA training conducted in Thailand in 1990 took six weeks, which was considered not long enough (Grandstaff *et al.*, 1990). PRA training in India has been much shorter. It has been hands-on participatory field training. A typical training experience has required camping in a village, and learning and using various methods. This is all part of a participatory process which leads to identifying actions by and with villagers. Staying a number of nights in the village intensifies and concentrates the experience. Villagers are encouraged to map, diagram, participate in transects, and plan. One aim of the training is to facilitate changes in outsiders' perceptions and behaviour, listening not lecturing, learning progressively, embracing error, being critically self-aware and participating themselves, for example by reversing roles and being taught by villagers how to perform village tasks. These experiences can open up a new range of possibilities and a sense of freedom to experiment and innovate. It is then not necessary to be trained in all the methods. Methods can be tried, improvized and subsequently adapted, and new ones can be invented. The creativity of both villager and outsider is released and developed through direct and mutual participation.

Potentials of RRA and PRA

The long-term potentials of RRA and PRA as they evolve are hard to gauge but look large.

In a more extractive, RRA mode, *ad hoc* investigations have been carried out on a myriad of topics. Examples include: why small farmers do and do not plant trees; why farmers do not maintain conservation terraces; how poor people spend lump sums of money; local practices of soil, water and nutrient conservation and concentration; how women and men spend their time; why an innovation has not been adopted; seasonal deprivation; migration; the impact of a road; the reality of what is happening in a government programme; the impact of structural adjustment on small farmers; and the rapid gathering of information for government decision-making.

Beyond RRA, the more participatory modes and methods of PRA have strengths. By transferring the initiative to rural people, PRA generates rapport and encourages outsiders to learn. By eliciting and cross-checking a lot of information in a short time, the process is cost-effective for outsiders as well as for villagers. By enabling people to present and analyse what they know, it generates commitment to action which may then be sustainable, as in Kenya and India. By being flexible and adaptable, and by encouraging inventiveness and innovation, it has the capacity to grow through the creativity of both rural people and outsiders. As it is often interesting, powerful and fun it engages all concerned, and makes them want to do more.

The longer term potential of PRA can already be indicted by practical applications. An illustrative list can include:

- Participatory watershed planning and management (including rapid catchment analysis (Pretty, 1990))
- Degraded forest assessment, protection, nurseries and planting
- Identification of credit needs, sources and interventions
- Health and nutrition assessments
- Planning the location of water supplies
- Assessments of biogas potentials and actions
- Selecting poor people for a programme, and deselecting the less poor
- Rehabilitation of small-scale gravity flow irrigation
- Preparing village resource management plans (NES *et al.*, undated)
- Participatory trials of crop varieties
- Identifying non-agricultural income-earning opportunities
- Investigating markets and smallholder marketing potentials
- Assessing and dealing with emergency situations
- Empowering women
- Orientation for students, NGO workers, government staff, and university and training institute staff towards a culture of open learning, and
- Participatory evaluation of programmes and planning the next phase.

And there have been many more.

Perhaps most significantly, PRA as it is evolving is an enabling approach for widely desired directions of change. It supports decentralization and diversity, providing means for local people to take command of their resources and to determine what fits their needs. By involving them from the very beginning of a development action, it helps them to own it more, and so contributes to commitment and sustainability. PRA is part of the paradigm for rural development which stresses process, participation, local knowledge and reversals of learning. Nothing in rural development is ever a panacea, and PRA faces problems of speed of spread, scale and quality assurance. But for the 1990s and beyond, its promise is evident. To make the 1990s a decade of local empowerment and diversity, participatory rural appraisal could have a key part to play.

Sustainable Small Farm Development – Frontiers in Participation

ROBERT CHAMBERS

Changing ideas in rural development

We live in an era of change unprecedented for its speed and unpredictability. In rural development ideas have been changing rapidly. Twenty years ago women were not much mentioned. Only 10 years ago, the environment was not high on the agenda. Now in 1991 we have an evolving consensus on the moving frontiers of development thinking and practice. It values indigenous technology, farmers participation in research, sustainability, and the enabling and empowering of rural people to gain for themselves much more of what they want and need. Increasingly, these changing values have been expressed in the concept of sustainable livelihoods, as a central objective that can be shared by rural people and by policy-makers.

Sustainable livelihoods

Livelihoods can be defined as adequate stocks and flows of food and cash to meet basic needs. Secure livelihood means the ability to meet contingencies without becoming permanently poorer. Sustainable refers to the maintenance or enhancement of resource productivity on a long-term basis. Rural livelihoods are diverse and often complex, with non-agricultural and non-farm as well as agricultural and farm sources. Nevertheless, the great majority depend upon production from a natural resource base.

Enabling rural people to gain adequate, secure and sustainable livelihoods would be a challenge if populations were static. With the population of sub-Saharan Africa growing at over three per cent a year (and liable at that rate to double in 20 to 25 years), the prospect is daunting.

Professionals as problems

The record of rural and agricultural development has been, at best, mixed. The question continues to be asked why performance has been so poor. Of all the changes of the past decade, one of the most hopeful has been the growing recognition that educated professionals are much of the problem, and rural people a large part of the solution.

This has been hard for us professionals to accept. Bookish education has given us the idea that our knowledge is superior. We know and they are ignorant. We should plan for them. Our packaged technology from research stations is good. It follows that those who do not adopt our recommendations, or who deviate from them, must be stupid and ignorant. But this explanation is no longer plausible. Today many are seeking other reasons why so many packages are not adopted. The answers are to be found not in them but in us.

Development professionals have often failed to understand small farmers' priorities, or why they do what they do. Isolated on research stations, researchers are often unaware of the nature of farmers' technology. From a detailed review of the literature, Reij (1990) concluded that 'our current knowledge of indigenous soil and water conservation techniques in Africa is extremely limited'. Professionals have rarely appreciated farmers' actions and abilities as domesticators and experimenters (Richards, 1985, Juma, 1989) or perceived the diversity and complexity of small farming systems.

While farmers manage and experience the whole of their farming systems, we are channelled by disciplinary training into narrow ruts. Soil and water conservation as a label does at least bring together soil and water, and professions which might otherwise consider soil and water separately. But our lack of multi-disciplinarity often leads to the neglect of much else, especially those dimensions which are social and economic.

Then, within our disciplines, we practise reductionism. We simplify complex reality into its parts, and control conditions in order to study and measure it. Our research, in consequence, generates standard packages from controlled environments. Centralized organizations then disseminate these as uniform solutions, but they fit neither the complex and diverse farming systems nor the unpredictable and uncontrollable conditions of farmer clients. In order to raise production and reduce risk, farmers, in contrast, seek not to simplify but to complicate, not to standardize but to diversify.

Our time frames also tend to be short. Projects are often to be completed in three or five years. In the field, targets require that physical works are completed by the end of the financial year. We need to learn from these mistakes, and see how we can do better in the future.

Farmers as part of the solution

The picture should not be exaggerated. There have been successes (Conroy and Litvinoff, 1988). And these often point to rural people, and especially farmers, as the source of solutions.

Farmers here refers to women and men, especially resource-poor smallholders. There are three respects in which farmers are a key to finding solutions: their knowledge, their time horizons, and their analytical capabilities.

Recognition of the validity and usefulness of indigenous technical knowledge is now widespread (Brokensha et al., 1980, Richards, 1985). Farmers are experts on most aspects of their farming systems, and especially those which are readily observable. They have to be in order to survive. Of course, scientists have knowledge, especially of microscopic phenomena, which farmers lack. But in terms of their farming practices, their priorities and their constraints, farmers have a comparative advantage.

The stereotype of poor farmers as always living from hand to mouth, and having no thought for the morrow, is not borne out by observation. It is true that those who are desperate will sacrifice the long-term for the short-term survival, and those who are insecure and fear loss or displacement, may not invest for the future. But there is much evidence – in planting and protecting trees, in indigenous soil, water and nutrient harvesting and concentration, in strategies to increase land holdings – that whenever they are secure, and even sometimes when they are not, small-farm families have a propensity to strive to invest their labour for future benefits.

Evidence has also accumulated that farmers have considerable ability to analyse. (Chambers *et al.*, 1989, Mascarenhas *et al.*, 1991). What has been missing is our ability to facilitate their analysis. If farmers analyse their farming systems they automatically screen out much redundant information. Recent participatory experience suggests that if rapport, methods and materials are right, farmers can analyse efficiently and produce sound decisions about development initiatives.

The case of deposition fields

The differences in the thinking, perception and priorities of farmers and of trained professionals can be illustrated with references to parts of north Karnataka, India, where farmers have for some decades been making deposition fields in *nallas* (seasonal watercourses). They build low stone barriers to trap silt, and gradually build these up year by year to make larger and deeper fields, fitting the structure to the landscape. The barriers are designed to collect soil, water and nutrients in these fields, to provide an important and relatively stable source of food and cash.

The government programme for soil and water conservation, in contrast, constructed check dams of a standard design. These were completed and then left. Typically, they were larger and higher than the farmers' silt trap barriers. They were designed for the professionals' objectives of slowing water flow and reducing erosion.

This case presents some common contrasts between professionals and farmers. The professionals took a short-term view, and built structures to meet their annual targets. Their general objective was to conserve, that is to keep soil in place and to slow and trap water. The farmers took a long-term view, to build up a productive resource. Their general objective was to concentrate, that is to trap soil, water and nutrients and build up a sustainable source of stable livelihood. For outside professionals, erosion is always an evil. For farmers it can also be bad. But sometimes, as in this case, it can provide a costless means of transporting soil to places where it will be more productive.

As this Indian example illustrates, the training and incentives of professionals condition them to perceive soil, water and agricultural production differently from farmers. Recognizing this is basic to participation. For unless it is farmers' priorities, rather than those of professionals, which are being met, farmers are unlikely to participate. Unless they participate, soil and water conservation is unlikely to be sustainable.

Participation

In the past five years, the frontiers of participation have been on the move, as more has been learnt about farmers' ability to analyse and innovate, and how outsiders can assist them. In rural research, approaches and methods have been and are being rapidly invented and evolved, but are still underdeveloped, very much a frontier.

Three streams of innovation stand out in participation: participatory rural appraisal (PRA); participatory technology development (PTD); and community participation.

PRA

PRA is a recent offshoot from rapid rural appraisal (RRA). Whereas RRA is extractive, with outsiders extracting information which they then take away and analyse, the term PRA can be used to describe a process which is much more participatory, with rural people presenting, analysing, owning and retaining information, but sharing much of it with outsiders. With PRA, development professionals act in a facilitating role.

PTD

There are many labels for farmers' participation in research and development (R & D) *Farmer-Back-To-Farmer* (Rhoades and Booth, 1982), *Farmer Participatory Research* (Farrington and Martin, 1988), *Participatory Technology Development* (ILEIA 1989), and *Farmer First* (Chambers et al., 1989). The labels do not matter – the substance does.

The essence of PTD is a shift from the transfer of technology (TOT) model. In TOT, technology is generated by professionals in research stations and laboratories, and then transferred as packages to farmers who are taught and trained. In the PTD model, teaching and training does still have some part to play, as with simple designs for farmers' own experiments (Bunch, 1985). But central to PTD are farmers' own analysis, design, observation and evaluation, conducted by themselves. The roles of scientists and extensionists then change. They are not transferers of technology, but convenors of farmers' groups, catalysts and facilitators of farmers' discussions and analysis, searchers for and suppliers of what farmers want and need, consultants for farmers' experiments, as well as tour operators who arrange visits for farmers to learn from each other. The main aim is not to transfer technology but to enhance farmers' competence, and assist them to identify acceptable opportunities.

In PTD, the standard package of practices of TOT is replaced by a basket of choices from which farmers can select for their diverse and complex farming conditions. A function of the formal research system is to help farmers to generate choices. Farmers will continue to experiment, innovate and adapt technology themselves, doing their own R & D, while development professionals facilitate these processes.

Community and group participation

The third stream of innovation is in relation to communities and interest groups. There are many traditions of community organization and participation in resource management (Chamala and Mortiss, 1990). What is perhaps most new is the awareness of the potential for group and community action to promote sustainable development. Fields of potential include:

- Managing resources held in common. This includes communal lands, forests, rivers, and bodies of water, the control of pests and diseases, and soil and water management where interests are linked, as in watersheds
- Managing funds raised locally, or provided by government or NGOs, and providing services such as credit
- Selection of farmers for trials, and as experimenters, monitoring and learning from their experiences, visiting innovators, and spreading good technology, and
- Making demands on government agencies, including demands on extension to search for needed information and genetic material, and on research for work on farmers' priority problems and opportunities.

Sustainable livelihoods

The three streams of innovation in participatory development intermesh, and each is constantly changing. In these newer forms, agricultural research and extension roles are reversed. Farmers do most of the survey and analysis, make requests and demands, and then experiment with their own R & D to develop technology with a local fit. Much of the significance of participatory approaches lies in their potential for generating sustainable livelihoods for future rural populations. Four aspects stand out.

The first concerns the intensification of farming systems. The association of production technology with population density is strong. In general, as population to land ratios rise, and as farm sizes decline, so farming systems are intensified. New enterprises are added, and internal linkages multiplied, both to increase production and to reduce risk. Farmers' comparative advantage in analysis compared with scientists' rises with the complexity of the system.

The second aspect concerns farmers' priorities. It is farmers who are the experts on their priorities, which can differ from those of scientists or officials (Chambers et al., 1989). Unless farmers can express their priorities, and through participation make demands on research and extension, the technology provided to them is liable to be inappropriate or harmful.

The third aspect concerns security and the long term. Secure land and tree tenure, and access to other resources, can be preconditions for farmers taking a long view and investing for sustainable livelihoods. Secure tenure can create a virtuous circle. The more secure farmers feel, the more labour they are inclined to invest for the long term. In turn, the more labour they invest, the more secure they feel.

The fourth aspect concerns dynamism and competence. Conditions are never static. Small farmers face changing and unpredictable physical, social and economic environments. To gain a sustainable livelihood, a farming family must be dynamic and innovative. Professionals have a role to play in enabling farming families to be alert and active, enhance their competence, and in promoting and supporting group and community action.

Conclusion: the first frontier

The conclusion is a paradox. Sustainable development is to be sought first not in the farming family, or the community, but in ourselves, the trained professionals. Our power, beliefs, reductionism and short time horizons are much of the problem. While farmers' knowledge, systems thinking, long-term investments, and enhanced competence and participation are much of the solution. It is not a case of either professionals' knowledge and competence, or farmers' knowledge and competence. The need is for a balanced mix, which means a shift to the farmers' side to enhance farmers' analysis and innovation.

For this balance and enhancement, recent experience points to the primacy of outsiders' behaviour and attitudes. By lecturing instead of listening, teaching instead of learning, by 'holding the stick' instead of handing it over, development professionals all over the world have inhibited the creativity and communication of rural people. The challenge now is to change our methods and reverse our behaviour. It is to enable and empower farming families to express their knowledge and strengthen their analysis. Thus, the role for new science-based physical and biological technologies is to provide baskets of options from which farmers can choose for local fit. In the 1990s there will continue to be traditional questions of how to generate those technologies. But more important will be the new frontier, of how to develop and spread among professionals effective means to enhance farmers' participation, competence and choice.

Participatory Rural Appraisal for Agroforestry

M. AVILA

Understanding farming systems

Agroforestry practices are ubiquitous, irrespective of agro-ecological and socio-economic environments. There is a very wide range of tree species and methods. As it is always easier to improve existing systems than to invent new ones, a good start for agroforestry research and development is to identify the types, roles and respective uses of woody species within the farm geography. Boundaries, home compounds, cropland, woodlots, pastures and other components of land use may be distinguished, and their tree components investigated. It is important to identify the order of importance in respect to:

- Use, for example fruits, fodder, medicines, fuel, timber
- Production practice, for example establishment, spatial arrangements, maintenance, harvesting and preparation for marketing, and
- Interrelations and competition between enterprises. Understanding the relationships, in the context of labour and capital between on-farm activities, is a prerequisite to an understanding of household strategies and tactics.

For the farmer, agroforestry represents an excellent strategy which allows:

- Exploitation of steep rocky hillsides and other marginal land types
- Diversification to capitalize on changing opportunities
- Use of multipurpose trees to provide a range of products
- Reduction of risk and uncertainty through a mixing of plant species and minimizing external inputs, and
- Satisfaction of a range of basic household needs, such as nutrition, cash, energy, construction materials, shelter.

Some of the farmers' practices to be examined in depth are:

- Perennial cropping and intercropping – such systems provide a permanent soil cover, improve infiltration and promote moisture retention
- Crop rotations – rainfall permitting, farmers use combinations of short and long season crops, including grass cover, to generate needed food and forages while protecting natural resources
- Crop residues – residues provide protective cover and are left on crop fields, except when farmers have livestock which compete for some types of crop residues
- Grass strips and hedges – particularly on sloping land, these are used to provide useful products, and
- Physical barriers – these would include terraces, stone-walling and other means of soil protection.

Identification of constraints

There are a few questions that must be asked in constraint analysis. These are, what are the priority needs of farmers and their households – food and nutrition, cash income, health, fuelwood, and so on? How can priorities be set and met through improvement of existing production systems? Should the development emphasis be on vegetable gardens, annual crops, perennial crops, fruit or timber trees? For each promising alternative, what are their respective biophysical or socio-economic constraints, or both, such as inappropriate species, low soil fertility, poor rainfall, labour bottlenecks, high input prices, and so on? Constraints could be divided into those emanating from within the farming system and those relating to the external environment.

Once the constraints are known and agreed by target farmers a matching of farmers' priorities with appropriate candidate agroforestry options, including improvements of those already practised, is a critical next step. A given problem can be addressed by several technologies, and one technology can address several problems. In some situations non-agroforestry solutions can have a greater comparative advantage. Often farmers have several equally urgent problems, and the best way to deal with them is through a combination of measures.

The appropriateness of a new or improved technology will depend on its biophysical feasibility and its socio-economic viability. Before assessments can be made, it is necessary to specify the details of proposed technologies in terms of:

- Basic problems of the household it is intended to tackle
- Appropriate species of trees, crops and animals
- Spatial and temporal arrangements with associated plants or animals, or both
- Source and preparation of planting materials, and their management, establishment, maintenance, production utilization/processing, and the marketing of products
- Required labour and commercial inputs, and
- Anticipated outputs, for example yields of main products in poor, average and good climatic seasons.

A salient conservation problem like soil erosion, which is a high priority from the community's point of view but perhaps not from the farmer's, may be addressed effectively by interventions whose main benefits are food security and cash generation. Today, much research work is in progress exploring the potentials of agroforestry for reducing soil erosion and improving soil fertility (Young, 1989, ICRAF, 1991).

Design of research and extension agenda

The specific questions raised during the technology design will suggest the need for on-farm research and experiment station research. On-farm research is an effective strategy to deal with problems related to farming system constraints, farmers' response and ability to manage specific technology designs, technology adaptation to local conditions, and definition of appropriate recommendation domains. Experiment station research is often more effective for basic research problems related to determining biophysical potentials of tree species, tree/crop/livestock interfaces and management tactics. But this research is site specific. It will not tell us how a particular species will perform at the target location. Both approaches to research should be complementary and feed into each other to achieve the overall objective of developing appropriate technologies which are adoptable by the target farmers. In the case of agroforestry, where basic and applied research are not well developed, and where farmers have much experience, on-farm research may have a stronger role to play in technology development than in crop or livestock research.

If technology testing progresses well during participatory technology development (PTD), and if farmers are anxious to try some designs, work could move rapidly into an extension mode, at least into a pilot phase of technology validation-cum-demonstration. This strategy can yield good results particularly when farmers take a leadership role with strong back-up from researchers and extensionists.

A combination of on-farm, and experiment station research and extension is ideal. Emphasis upon on-farm research initially should result in early benefits to, and stronger participation from, farmers. Benefits from supporting experiment station research and extension should lead to further improvements in land husbandry later.

A Multi-disciplinary Approach to Socio-economic Constraints and Research Priorities

A.J. SUTHERLAND and L.P. SINGOGO

Introduction

Farming systems research (FSR) in Zambia has been implemented for almost a decade, but land and water management (LWM) issues and problems have been neglected. The nation has reawakened to the need for soil and water conservation. Today, fertilizer prices are beyond the reach of many farmers. Also, continuous applications in conjunction with maize monocropping is resulting in soil acidification. Land with low slope is now scarce, and farmers are clearing and farming hillsides, with accelerated soil erosion as an observed consequence. Problems of declining soil fertility and soil loss by erosion are evident in the densely populated Eastern Province.

The institutional setting

Soil and water management (SWM) issues involve all three branches of the Department of Agriculture. The Research Branch conducts research. Irrigation and Land Husbandry Branch translates technical research data into LWM messages to be relayed to farmers through the Extension Branch network of field officers.

Within the Research Branch, adaptive research planning teams (ARPT) conduct farming systems research. Each provincial team comprises an agronomist, an economist, and a research-extension liaison officer (RELO). Some have livestock specialists. In addition, rural sociologists provide an input at regional level, while at national level there is a nutrition co-ordinator and a national co-ordinator. Soils research at national level is conducted by a specialist research team. There is no focus upon water research and crop research is dealt with by commodity teams.

Public sector realization of land degradation in Eastern Province

A World Bank funded agricultural development project started FSR in 1982. The work emphasized identification of production constraints and the development of technical recommendations related to cultural practices and improved varieties for the major crops in the farming system of the province. Using the CIMMYT (1980) approach to problem identification and priority setting the provincial adaptive research planning team concentrated on technologies which promised short-term returns, such as new varieties, inorganic fertilizers, and related cultural practices such as planting dates, weeding methods, timing and placement of fertilizer.

Awareness and concern about conservation issues came after about four years, and in 1986 two exploratory trials for the establishment of grass banks on steep slopes and improved fallowing were initiated. Due to staff changes and limited technical support this research work was not sustained. However, interest in conservation increased through activities of the Soil Conservation and Agroforestry Extension Project (SCAFE). As part of this project, ICRAF conducted in-country training on diagnosis and design methodology for extension staff in 1987. This methodology was later used in the districts to identify soil conservation problems and develop relevant extension activities. These developments provided a fertile ground for a fuller involvement of the ARPT in conservation research.

The restoration of fallow and degraded lands project

This project, initiated by ARPT, aims at addressing problems in Eastern Province through a phased multi-disciplinary approach. Phase one involves problem identification and characterization. Phase two comprises screening of possible solutions through on-station and on-farm experimentation. In phase three promising solutions are developed into recommendations and tested over wider areas. So far the project has initiated phase 1.

A meeting of scientists

In order to initiate phase one of the project a multi-disciplinary team of 14 officers was assembled and a meeting convened. The objective of the meeting was to review the project proposal and develop a plan of activities involving all those present. Project objectives were emphasized. The meeting also generated a timetable, a plan of activities and allocated responsibilities.

Activities for phase 1

Activities identified were:

- A literature review to identify types and location of land degradation
- Informal survey of the land degradation in selected target areas to identify the major types of soil degradation, causes, and possible solutions

- Prioritization of research issues, and formulation of research strategies, and
- Area soil characterization and detailed socio-economic data collection.

The implemented activities

Literature review
The literature review was presented at a first meeting of team members on the first day of the survey. In retrospect, the development of sound terms of reference and naming the activity background data collection might have improved the situation.

Area selection
During the pre-survey meeting two points of view emerged. One group proposed a rapid tour to observe the full range of soil degradation problems and farming systems in the province, arguing that a good overview was necessary before selection of sites. The second group wished to study degradation in one farming system type. They argued for an approach in depth and detail, pointing out that with only 12 days for the survey, including analysis of data and drafting of the report, the proposed rapid tour could not be accomplished. The focus agreed was upon the most widespread and densely populated farming system/agro-ecological sub-zone. The range of variation within that system would be covered by selecting three sample points, each in a different district.

It was agreed that, within the districts, area selection would be made by the team based on information provided by the local agricultural extension staff. The same procedure would be used for the selection of villages and farmers to be interviewed within the areas selected.

The survey timetable
The timetable for the informal survey was agreed as Katete District three nights, Chadiza three nights, Petauke four nights, Chipata three nights (for analysis and write-up). This was a departure from the provisional timetable and meant that various district agricultural offices had to be informed of the changes. Not all the changes were effectively communicated leading to inconveniences on both sides.

The second team meeting

Attendance differed from the first meeting. Of the original 14, eight were present and five new members were added, but one of these dropped out after the meeting due to dissatisfaction with the accommodation and subsistence arrangements. The team was further reduced in size when the soil surveyor and the economist were called away for other duties during part of the survey. The most conspicuous absentees were the agroforesters who were attending a meeting out of the country at the time.

Survey methods and approach
The day before the survey commenced team members were given a short article to read explaining the informal survey method to be used (Rhodes, 1985). Before starting the interviews, the team had a briefing on informal survey methods and quickly reviewed a checklist of issues to be covered during the survey.

The first day of interviews was primarily a training exercise. It started with an interview with the district agricultural officer (DAO), followed by an interview with a field extension worker and then a farmer. After each interview a brief review meeting was held to discuss mistakes made during the interview. At the end of the day the team met to review progress and make plans for the next day.

The next day the team subdivided and went to different parts of the district accompanied by staff from the DAO's office. Towards the end of the day another review meeting was held, and the following day further farmer interviews were conducted in the two areas, including group interviews. Soil samples were taken concurrently with interviews which were conducted on problem sites in farmers' fields. At the end of the interview period in each district a summary of land degradation problems, causes and indicators was made. This procedure was repeated in the other two districts.

A further stage of analysis began on arrival in the last district, with the listing and screening of potential solutions to problems identified. Some of these solutions were then presented to farmers during interviews in order to get their reaction.

After the findings for each district had been summarized, the team began to compare similarities and differences between the areas, and generate tentative hypotheses about the probable causes of these variations. In addition, problems and potential solutions were ranked, respectively in terms of importance and potential for adoption. The team was then subdivided into smaller groups responsible for writing up different sections of the report with deadlines for the following day. On the last day the different sections for the report were submitted. One local member of the team was given the task of compiling and editing the report and ensuring that outstanding bits of information gathering and writing were undertaken by those responsible.

Problems identified

The main problems identified during the survey were:

- Soil erosion on both arable and grazing land
- Food insecurity
- Fertilizer inaccessibility and dependence
- Shortage of draught power and ox-drawn implements, and
- Labour constraints during critical periods.

Socio-economic factors
It was observed that land degradation problems were meshed with socio-economic factors. The nature and extent of land degradation was closely

linked to the history of human settlement and population movement. Soil fertility problems were most acute in Chadiza, the most densely populated area characterized by sandy soils, flat terrain, regular rainfall and healthy conditions for cattle. Over the past two decades a gradual movement of people westward has occurred towards heavier soils, hilly terrain, less regular rainfall and more cattle disease.

Land tenure is also an important factor. Uncontrolled communal grazing inhibits control of cattle to minimize their contribution to degradation processes. Arable land is held by families. Some families with larger holdings and greater security were able to leave areas fallow. Others with smaller holdings, and perhaps less security, felt obliged to cultivate their entire fields, in order to prevent others staking claims to the uncultivated land. This results in poor management of the planted crop and very little free labour to implement soil conservation measures, such as construction of level bunds or increasing the area planted to leguminous crops.

The effects of the larger economy also had an impact through increased fertilizer and seed prices. Farmers complained that they now had to plant larger areas of local maize in order to feed themselves, but had reduced their areas of high yielding hybrid maize because they could not afford to buy the seed and fertilizer at the new prices. To reduce their costs some have started using cattle manure. However, this source of fertility is available to the wealthier minority. The preliminary evidence indicates that the majority of small-scale farmers use neither manure nor inorganic fertilizer.

Potential solutions

Technical solutions proposed by the team were scored for potential adoptability by the team. Those identified with potential are: galauza, alternative varieties, intercropping, wetland cultivation, fertilizer/lime inputs, contour farming, crop rotation, animal manure inputs, alternative crops, tree planting and supplementary staples. Solutions with low adoption potential were identified as: water harvesting, commercial organic fertilizer, compost manure, cover cropping, controlled grazing, first rains land preparation, alley cropping, fallowing, winter ploughing, improved pasture, soil conditioner, and mulching.

Galauza is a traditional method of cultivation. After harvest, stover and trash are laid between the crop ridges. Ridges are then hoed out to bury the stover and create new ridges in the previous furrow. The other most promising solutions follow the trend of pre-independence research and extension activities, or are concurrent with farmer strategies and innovations.

The next step in pre-screening potential solutions will be to hold further discussions with the farmers in order to establish whether the teams' ranking of adoptability corresponds with the farmers' opinion.

Lessons learned

The informal survey proved to be a fast and efficient way of identifying land degradation and related problems. In-depth understanding may have increased if more time was spent in fewer areas. Socio-economic issues were investigated superficially, but the identified problems were made known to all members of the team. Certain aspects of land tenure, access to animal draught and manure, labour constraints and food security were not investigated. But these factors are likely to be critical to a full understanding of problems and potential solutions. For the future an in-depth focus on socio-economic studies is likely to be advantageous.

Multi-disciplinary teamwork
In general, the combination of different disciplines worked out well and provided a healthy breadth to the range of issues covered during the survey. Subdivision of the team into small groups (two to three a group) during the survey enabled the different disciplines to be spread out. Discipline balance in sub-teams was impossible because of the need to allocate members with knowledge of the local language. Inevitably, this resulted in certain biases which were reflected in the type of questions asked and data collected. The daily post-interview meetings served to bring the different types of information together and allowed the team to set the general direction of questioning for the following day. These meetings were crucial, although strenuous and tedious to some team members. Without them the whole exercise would have probably failed.

Co-ordination and leadership
The whole exercise was quite taxing in terms of co-ordination and leadership. Different financial arrangements for different members of the team, adjustments to the pre-arranged timetable, difficulties in finding accommodation, co-ordination of transport and difficult eating and sleeping conditions had a major effect on morale. Most complaints came from the younger and less experienced team members, who also showed the earliest signs of fatigue. However, breaking up into smaller groups with specific responsibilities proved to be an effective way of keeping everyone active and motivated. The long evening meetings proved the most difficult to co-ordinate. Rotation of discussion leaders and secretaries proved helpful in maintaining interest and spreading a sense of responsibility across the group.

Institutional issues
The survey brought together professionals with different geographical and institutional allegiances. Half of the team came from the national headquarters. The other half came from within Eastern Province. All were nationals, except for two expatriates. The Research and Irrigation and Land Husbandry Branches were both represented. One team member was from the University of Zambia. Within the Research Branch, different teams and sections were involved.

The fact that the exercise was initiated by the farming systems team may have accounted for the participation of senior scientists in that team, while the soils research team was mainly represented by more junior scientists. This situation called for leadership skills by the ARPT national co-ordinator, who also managed to have sufficient finances to cover shortfalls in respect to non ARPT staff. Had each section participating been depending on finance from their section the operation would never have been implemented. The initial intention to give leadership responsibility to the local team did not work well in practice and the ARPT national co-ordinator had to take over after a short period to avert divisive internal wrangles.

The exercise also made clear that the current research organizational structure, budgeting, and funding is not conducive to research programmes which demand the involvement of multi-disciplinary teams. While the branches of the Department of Agriculture agree on the importance of addressing sustainable agriculture issues through a multi-disciplinary approach, little effort has been put in to make this a practical reality.

Opportunities and Benefits

Introduction

N.W. HUDSON

Introduction

There are three groups of case studies. In this section we look first at the opportunities to improve development projects by learning from surveys of earlier projects and of farmers' attitudes. This is followed by case studies of four examples of looking forward to improving the situation. The next section includes nine case studies where the emphasis is on helping communities to develop their abilities and responsibilities, and the last group of case studies focus on improving land husbandry in site-specific cases.

We have reviews of projects in Kenya and Tanzania, and surveys of farmers' attitudes in Uganda and Ethiopia. The three projects in Baringo reviewed by Dr Asenath Omwega and her co-authors are all 10 or more years old, and mainly illustrate how much we can improve on early projects. Baringo is a harsh environment, semi-arid with poor degraded soils, and all the familiar problems of much of eastern and southern Africa. The primary cause of the poor results of all three projects was the lack of farmers' participation. They were all brought in as externally designed projects, and attempts to involve the local people after the project was set up did not catch on. They did not mobilize the strong support for voluntary labour which exists in Kenya under the slogan *harambee*, which loosely means, let us all work together.

Another weakness which project initiators are unlikely to repeat nowadays was that none of the programmes paid enough attention to women's role in the farming system. Kenya has made good progress towards decentralizing authority to the districts, so the problem of integrating project initiatives and the work of line departments need not have arisen. It was another result of bringing in a project already designed somewhere else.

The question of how to arrive at a coherent plan to integrate a new project with existing line departments was also addressed in both of the reviewed projects in Tanzania, SECAP, and HASHI. One of the main reasons for the poor result of HASHI was the poor organizational structure. There were inadequate links from central government to district. The contact ends at the district level, as there is no village project management team. This led to inefficient allocation of resources between the various project components because of the lack of flexibility. The overall result was a low level of involvement of the local peoples, and a poor record of implementation.

SECAP has, like the Baringo project, been operating for more than 10 years. However, it has had the advantage of sufficient flexibility to react to difficulties and to modify the approach. For example, initially SECAP operated with extension staff seconded to the project who dealt with SECAP activities only. But as the programme broadened, a joint extension body was formed under the district extension office.

The main thrust of the project has been targeted on livestock management and this has been successful through two main themes – improving the local cattle by cross-breeding with imported bulls, and improving the fodder supply through macro-contour lines, which are two-metre strips of high yielding fodder plants laid out on the contour through the arable lands. Later, after the basic concept of macro-contour lines for fodder production was accepted by farmers, the use was broadened to include food crops such as pineapples, bananas, and sugar cane.

The other components of the project were less successful than the livestock improvement, and the project has not yet overcome the problem bf adjusting the extension package to the needs of women farmers, non-livestock farmers, and to strengthen the soil and water conservation. This might appear to be an argument for projects with a limited attainable objective, and there are other examples of successful single-focus projects, for example fuelwood. There can also be a case for moving one step at a time as suggested by Shaxson in his paper on a strategy for Thabana Morena in Lesotho. However, the weight of opinion of project operators seems to be in favour of aiming for development across the whole of the farming system. This supports the contention of the authors of the HASHI review, that each of the project components must be adequately tested on-farm to make sure that it is acceptable.

The surveys of farmer attitudes in Uganda by Mrs Joy Tukahirwa, and in Ethiopia by Yohannes G. Michael shows some interesting similarities. In both there was a strong awareness of the problems of soil degradation, but a significant reluctance, sometimes antagonism, to doing something about it.

In Uganda there is a residual historic objection to the imposed soil conservation measures of colonial times, and in Ethiopia to the dictatorial attitude of the socialist government through its devolved party system which took political control right down to the village level through Peasant Associations. Clearly it is necessary to overcome this opposition and establish a political environment which can stimulate and encourage rural development. Kenya had the same legacy of colonial conservation but has largely been able to overcome it, as have Zambia and Zimbabwe.

Some other aspects of the reluctance to adopt conservation farming common to both these surveys of farmers' attitudes are that poverty and malnutrition dictate the need to seize any short-term profits, even when it is realized that these are at the expense of long-term resources. Another problem which comes out strongly in both these surveys, and indeed in a very large number of the case studies, is the question of lack of security of tenure to cultivation rights, or grazing rights, or ownership of planted trees. In addition, there is an almost universal shortage of land and a shortage of labour occurs as the menfolk are increasingly forced to seek employment off the farm.

The Ethiopia survey shows that the farmers make a strong link between wealth and the ownership of oxen. This relation is not only made on the basis of social status, but also because the availability of oxen for ploughing and cultivating at the right time has a huge effect on productivity.

The paper details an elaborate social structure of mutual help arrangements, some of which are: *Zeye*, where two farmers having a single ox make a pair to plough the land in turn; *Degwa*, where a farmer with a pair of oxen makes a temporary agreement to plough the land of a poor farmer in return for an equal share of the crop; and *Shufa*, which is a co-operative saving scheme where a group of 10 or more farmers each contribute an amount of money weekly with a pay-out to a member facing a serious problem, such as payment of taxes or the purchase of an animal.

Increasing population is reported to be recognized as a problem in most countries, and the Ethiopia study looked at this question in some detail, and reported:

Most of the farmers said that they had a big family and face problems of feeding and clothing them. On the other hand, they have a strong wish to have more children and the following reasons were given:

- Providing additional labour for the farming activity
- Having additional income from off-farm activities of the children
- To get help in old age
- Insurance against the frequent death of children
- Social status to have a large and strong kinship, and
- It is a gift of God.

The decision to have more children appears to be both an economic and a religious issue. About thirty-seven per cent of the farmers have two or more wives and the following reasons were given: 'Mother may be sterile so take another wife'; 'Mother may give birth to two or more females, so to get another son take another wife'; and 'It is an indication of wealth'. The farmers were asked if they had any advice on family planning and replied that they had not and are not interested to know about it, since it is 'against the will of God'.

The report goes on to make recommendations to minimize the population pressure. These are:

- Introduce techniques which reduced labour requirement
- Strengthen the local labour-oriented institutions to use labour more efficiently and effectively
- Introduce a cottage industry to absorb the excess manpower, and
- Introduce family planning advice, using religious leaders who command much respect in society.

It is clear that the Ethiopian farmer has a good understanding of the nature of his or her problems and their causes and effects, but is unable to see any way out of difficulties. It is interesting to compare two personal stories, one from Ethiopia and one from Kenya.

The Ethiopian farmer is between 65 and 70 years old and describes his situation as follows:

I was the only son of my father, by chance the others were females. When my father died, I possess his big land. I was working hard and getting rich and richer. The homestead field was very wide, full of enset, coffee, sugar cane and banana. Production was very high since I used manure and sometimes fertilizer. Downstream, the grassland and woodland was very wide. I had enough livestock, including three oxen. Gradually I married a second wife, later on a third wife, both of us are living in the same compound but at different houses. Now the total size of our family is about 14. The elder son is married and shares part of my land. The Peasant Association took part of my land and assign to a landless farmer. As you can see then, most of my children are kids. There was one young unmarried boy aged 17 who used to help me a lot but the PA took him to serve as militia since the last two years. I could not feed and clothe my family and at the same time pay taxation. So I started selling my livestock. Some died because of disease. Now I have no more livestock, no more manure or fertilizer, made share cropping agreement, cultivation on all the grassland and part of my woodland. Still I could not solve the problem, and do not know how, gone are the good days.

The extension worker who accompanied the interviewer asked 'how can I tell this man about land management? How can I tell him to think of his future, when he is not sure of surviving today?'

Compare this depressing story with that of the peasant farmer in Kenya described as John Kagwe in the paper later in this section by Cheatle and Njoroge. John inherited a smallholding which seemed incapable of supporting his family. One third of the land was under tea with a big labour demand and poor cash return. Food was insufficient, and a loan to buy fertilizer did not solve the problem. Servicing the loan, paying medical bills and sending a child to school used up the cashflow.

By 1987 John regarded himself as near bankrupt and believed himself in danger of losing his land. He described: 'there was not enough money, not enough food. My wife run away'. In desperation he took a training course for organic farmers and worked out a new farming system. The tea was rented out and from this income he was able to pay off his loan, buy some small stock and set up as an organic farmer using the improved farming methods described in detail in that paper.

After two or three years he had managed to climb out of the trough and reported, 'my wife is back now, we have more money, more food and are healthier. We send five children to primary school, even though we have to pay school fees and for books lunches and uniforms'.

So, there are seeds of hope. Even apparently impossible situations can be retrieved.

The papers summarized are: Omwega, A.K., Ngunjiri, M., Njoroge, S.N. and Obando, J. 'Evaluation of soil and water management projects in Baringo District of Kenya'. The three projects reviewed are: BPSAAP Baringo Pilot Semi-arid Area Project, sponsored by World Bank and Government of Kenya; BFAEP Baringo Fuelwood Afforestation Extension Project, supported by FAO and the Government of Kenya; and BFFP Baringo Fuel and Fodder Project supported by the Netherland Government and the Government of Kenya.
Rugumamu, W. and Rwejuna, C.S., 'Environment programmes for sustainable smallholder development in Tanzania: a case of HASHI'. HASHI stands for *Hifadhi Ardhi Shinyanga*, which means land conservation in Shinyanga.

Sianga, C.W. and Diederichsen, H., 'Towards restoring the environment: the SECAP experience in Tanzania'. SECAP is the Soil Erosion Control and Agroforestry Project sponsored by GTZ and the Ministry of Agriculture, Livestock Development, and Co-operatives.
Tukahirwa, J.M., 'Attitudes to soil conservation among smallholders in Kabale District of Uganda'.
Yohannes G/Michael, 'Indigenous farming systems in Wolayta, southern Ethiopia'.
Requests for further information on these studies may be addressed to the authors.

Choosing Conservation Measures for Cropland on Smallholdings in Kenya

D.B. THOMAS

Introduction

Conservation of soil and water depends a great deal on crop and soil management which are part of normal farm operations. The management strategies that lead to good yields will usually be the best for conservation purposes (Hudson, 1981). Conservation should be an integral part of the farming system, and conservation measures should increase production or at least halt a decline.

The best conservation measures are associated with improved soil and water status or improved fertility. They are sometimes accompanied by a change to more intensive systems of farming with higher value crops or irrigation. In semi-arid areas and on poor soils, the emphasis should be on water conservation and maintenance of fertility, rather than on protection of soil. Conservation of rainwater *in situ* will minimize the loss of soil. Measures which improve plant cover will normally lead to greater infiltration and protect the soil from raindrop impact.

There are some conservation measures which cannot be expected to lead to improved yields. But these may still be essential to control runoff and reduce damage from gullying or sedimentation. These include cut-off drains, artificial waterways, and structures for gully control.

As conservation is carried out by farmers on their own land, they must be involved in planning and evaluation. The farmer and extension worker should be assisted by the conservation scientist who has access to many sources of information and can carry out research where it is lacking.

In the past conservation programmes suffered from blanket recommendations which were not closely tuned to the local situation. Action was hampered by lack of scientific knowledge and understanding as much as by farmer inertia. The scientific approach to conservation is still poorly developed and there are many unanswered questions. However, from the experience gained, it is possible to be more systematic in analysing the options available, identifying the most appropriate for a given situation, and monitoring progress.

Analysis of options

The following steps are suggested in order to prepare recommendations for a given situation. They should be carried out in consultations with farmers.

1. List the possible measures and define them in such a way that they can be clearly identified and properly implemented.

Both agronomic and structural measures should be clearly defined. There is a good deal of confusion about terminology, for example, the words bund and terrace are often used interchangeably. Sometimes they mean the same thing, sometimes they do not. Again, there are many different ways in which they can be implemented. Some are satisfactory, some are not. Generally, there are no standards or specifications. If we think that agroforestry should be encouraged we should be very clear which form is most suitable and how it should be implemented. Practised in certain ways it may be beneficial, while in other ways it may not.

2. Assess the technical effectiveness of alternative measures to reduce on-site and off-site damage or maintain fertility, or both. In the past, the main emphasis was on protection of soil *per se*. Now there is greater recognition of the importance of water and nutrient availability. Some measures may have no direct impact within the field but may be important in reducing off-site damage. The impact of each measure should be assessed on the following criteria shown in the table below, to which a numerical rating could be added.

On-site contribution				
Erosion control	nil	minor	major	total
Runoff control	nil	minor	major	total
Nutrient level improvement	nil	minor	major	—
Soil structure improvement	nil	minor	major	—
Farmability increase	nil	minor	major	—
Off-site contribution				
Gully prevention	nil	minor	major	total
Sediment reduction	nil	minor	major	total

3. Determine the suitability of alternative measures for the specific conditions of climate, soil, slope, and land use.

Climate The main question to consider is the crop water requirements and the control of runoff. In humid areas where rainfall exceeds crop requirements, measures are needed for the safe disposal of runoff, otherwise waterlogging or excessive leaching will occur, or both. In semi-arid areas rainfall is often deficient, and conservation measures should retain rainfall *in situ*. In arid areas, rainfall alone is inadequate and should be supplemented by water harvesting methods.

The distribution of rainfall is also important. Where the rainy season starts abruptly and heavy rains may occur early, and when crops are barely established, the potential for erosion is very high. Where the rainy season starts slowly and the heaviest rains occur when the crop is well grown the problems of erosion are much less serious.

Soil The selection of conservation measures will depend on the nutrient status, erodibility, depth, permeability, stoniness, susceptibility to cracking, and liability of mass movement.

Slope The risk of erosion increases exponentially with slope. Slope does not fall into easily defined categories, but measures which may be adequate on low slopes may be quite inadequate on steep slopes.

Land use The type of land use has major implications for conservation. Where land is used mainly for perennial crops, the problems of conservation are fairly easy to solve. Where land is under annual crops the problems are much greater. Densely populated land presents special problems. Waterways for disposal of runoff may be unavailable and footpaths may be prone to gullying.

4. Identify the constraints which may prevent adoption. Too little attention has been given to the constraints that the farmer is facing. The following are some of those which need to be considered.

Land tenure Where there is insecurity of land tenure or the user has no long-term rights it is unlikely that conservation will be taken seriously. In some circumstances communal ownership may provide adequate guarantees to warrant investment in conservation, but often it does not. The greatest progress has been made in areas where there is individual title to land.

Size of holding Conservation measures may involve a loss of land for crops, and if land is short and there is no compensatory gain in fodder, fuelwood, or fruit, the measure proposed may be unacceptable. The value of fodder produced depends on the quality and importance of livestock in the farm economy. Where there are high-yielding dairy cows and a demand for milk, it is easier to justify converting land to fodder.

Competition with crop Certain conservation measures may lead to competition with the crop for water, nutrients, or light, and may reduce crop yields. Some systems of agroforestry and certain fodder grasses can be competitive for water and reduce yields in semi-arid areas, especially in seasons of low rainfall. Crop residues can be useful in reducing soil losses, but if incorporated in the soil they may lower the availability of nitrogen to the crop.

Weeds Weeds are a major problem for farmers, and grass weeds such as Kikuyu grass (*Pennisetum clandestinum*) or couch grass (*Digitaria scalarum*) which may spread from terrace embankments or from grass strips, are particularly troublesome.

Pests and diseases Retention of crop residues as mulch may lead to an increase in problems such as stalk-borer in maize, or glume-blotch in wheat. Grass strips can harbour rodents which are particularly troublesome in some areas.

Labour Shortage of labour is a constraint on some farms at certain times, and on some farms all the time. Lack of cash to employ labour prevents some land owners from implementing conservation measures.

Conservatism Traditional ways of farming have developed over centuries. Changes may involve risks. The need to adopt new practices may not be perceived unless there is a crisis situation, such as may be caused by population pressure, falling yields or drought.

5. Evaluate the costs and benefits of different measures in socio-economic terms. This is another area which has received inadequate attention and it is only possible to mention briefly some of the issues which should be considered. Although Pagiola (1990) has tried to analyse the economics of conservation in Kenya, some of the costs and the benefits are hard to quantify without more data. For example, there is little data to show the impact of soil loss from different soils on crop yield.

The costs of conservation measures may include:
- Crop production lost from land occupied by grass strips, terraces, hedgerows
- Crop production lost due to increase of weeds, pests, or diseases
- Labour required for implementation or construction
- Labour required for maintenance
- Materials required for construction, for example of stone terraces, and
- Crop residues left *in situ* which might have been used for fodder, bedding, or fuel.

The benefits of conservation may include:
- Fodder, fuelwood, fruit, or fertility gained from land occupied by grass strips, terraces, hedgerows
- Crop production gained due to rainfall retention, and
- Crop production gained by halting a yield decline, or improving nutrient availability.

Assigning an economic value to these items is not straightforward. For example, if family labour is used there may, or may not, be an opportunity cost of alternative employment which is foregone in order to carry out conservation work. If there is no alternative work and if the conservation can be carried out by the family during a slack period there may be no economic cost to the labour used. Where land is individually owned and there is a market in land, the improvements may be reflected in increased value added to the land. Under communal ownership no such benefit may accrue.

6. Determine farmer preferences and list adoptable measures. In any given situation there may be one or more measures which could be used. Sometimes the best solution is to use two or more measures in combination. The most effective measure is not always the most preferred. For example, on sloping land in a dry area we may decide that level bench terraces would be an effective measure because of the benefits gained from water retention. However, the labour required may be beyond the farmer's resources and a *fanya juu* terrace, which is a first step towards bench terracing, but requires less labour, may be more easily adopted. Alternatively, we may find that the farmer is not able or willing to invest labour in terracing but would accept narrow grass strips, as a step towards better conservation.

We could therefore list the options in terms of effectiveness in controlling soil and water losses, and then list them in order of farmer preference, as shown in the table below.

	Effectiveness	*Farmer's preference* (adoptability)
First	Level bench terrace	Narrow grass strip
Second	*Fanya juu* terrace	*Fanya juu* terrace
Third	Narrow grass strip	Level bench terrace

Farmers are not all the same. Where the order of effectiveness coincides with the preference of the farmer, rapid progress will be made. Where it does not, the farmer's preference has to be followed. Any progress is better than none. But the obstacles to adoption of more effective measures should be investigated. Conservation measures which are part of seasonal farm operation such as ridging, the use of crop residue mulch, compost, or green manures, should be carried out regularly to be effective.

Some structural conservation measures can provide good control of soil loss and runoff. If correctly installed they are permanent and maintenance is minimal. Other measures provide partial control and need regular and substantial maintenance without which they will not be effective and may disappear completely. Wherever possible the farmer should be encouraged to adopt permanent measures and avoid the risks associated with partial solutions.

Recommendations are not fixed once and for all. They need to be re-examined periodically, and revised as necessary, in the light of farmer experience and the results of research.

7. Define the procedures for monitoring the extent and rate of adoption of the most suitable measures. A question often asked is whether the rate of soil erosion is increasing or decreasing as a result of conservation programmes. This is not easy to answer. Measuring sediment output in streams and rivers is a difficult undertaking. The largest amounts come during big storms, usually at night and automatic equipment is needed for sampling suspended load. Bed load is even more difficult to measure. The rate of soil erosion on cropland may be falling, while the rate of soil erosion from road construction may be increasing. The result may be little overall change. Even if there is less sediment reaching a river from farmland there may still be large quantities taken up from the banks and bed.

An alternative approach is to assess the effectiveness of different conservation measures using rainfall simulators and runoff plots, and then measure the extent of adoption of the best practices. There is a good deal of knowledge, from local experiments and work elsewhere, about the benefits of different agronomic practices and conservation structures in controlling losses of soil and water. What is needed is to find out the rate at which farmers are adopting them. This is a survey exercise and sampling procedures can be devised which will give conclusions that are as reliable as we wish, or can afford.

A note of caution is needed here because the way in which a given measure is implemented can make all the difference between success and failure. We need to look at quality, as well as quantity, when carrying out an evaluation.

Survey methods which can be used include air-photo survey and ground survey. Air survey enables a large area to be covered and if the scale is about 1:12 500, a great deal of information can be obtained. Ground survey gives more detail, is easier to repeat and allows contract with, and comments from, the farmers. Because of the difficulty and cost of monitoring enough sample area to represent, say, a whole district, it may be better to select sites which are representative, and monitor the same areas every few years by ground survey.

One way to do this would be to use predetermined transects from ridge top to valley floor, and carry out the survey at the time when most erosion occurs. This is usually during the early stage of crop establishment.

Developing a satisfactory method of evaluation is an essential step without which it is impossible to monitor progress.

Appendix: a summary of conservation measures in Kenya

Agronomic measures

Crop selection and management
Farmers rarely change their cropping practices purely for conservation purposes. Changes which are made for other reasons may have positive or

negative implications for conservation. The move from maize growing to tea, coffee, or macadamia that has occurred in parts of Kenya, has led to improved cover and reduced erosion hazard. The destruction of coffee, on account of low prices, would increase the hazard.

Early planting to provide a quick cover is important. Beans cover the ground faster than maize, and intercropping of maize with beans or cowpeas is much more effective in controlling erosion than growing maize alone.

Contour cultivation and planting
Where oxen or tractors are used, cultivation should be on the contour. Planting should also be on the contour so that if weeding is done by hand, ridges can be built along the rows.

Contour strip cropping
Contour strip cropping is practised in the USA, with alternating strips of cereal crop and pasture, so that the grass traps soil washed from the cereal crop. This is uncommon in Kenya, but a somewhat similar system does occur on some farms where strips of cereals alternate in a rather irregular manner with sweet potatoes or fodder grasses, or both.

Crop residue mulch and trash lines
Plant residues can play an important role in protecting the ground during the time of crop establishment. They are more effective if they are left on the surface than if they are buried. Burial can also lead to nitrogen shortage.

Contour ridging
Contour ridging is not a common practice in Kenya, except for potatoes, though it is almost universal in Malawi. Ridging can make a significant contribution to soil and water conservation on low slopes. Support measures, such as terraces or grass strips are needed in case ridges break during heavy storms. Tied-ridging has a lot to commend it in dry areas, but is rarely practised because of the labour involved.

Manure and compost
Manure is still not given the attention it deserves. There are large losses of nitrogen and other nutrients through ammonification and leaching, due to lack of bedding, or lack of shelter. The proper conservation and utilization of manure should be a high priority in view of the importance of organic matter and the high cost of fertilizers.

Some farmers have realized the value of compost and have been using manure in combination with wood ash, grass, chopped bushes and crop residue to make compost. Using manure in this way allows it to mature before application, reduces losses of nutrients, and increases the quantity of organic matter available. Examples of the successful use of compost are described in the paper by Cheatle and Njoroge in this section.

Green manuring
Green manuring is not a practice that has attracted much attention. The arguments against it are that there is insufficient land to withdraw a portion from other crops, and the effort of growing a crop simply to improve the soil would be unacceptable. But ideas are changing, and there may be a role for green manuring in some places with crops such as velvet bean (*Stizolobium sp.*), dolichos bean (*Dolichos lablab*), or sannhemp (*Crotlaria juncea*). Velvet beans have become popular in Honduras for this purpose, and sannhemp has been widely grown in central Africa and India.

Vegetative barriers
Vegetative barriers impede runoff and lead to the deposition of sediment. Narrow grass strips can play a useful role in controlling erosion and providing fodder for livestock without taking too much land. Where land is in individual ownership and productive stock are being stall-fed, high quality fodder such as napier (*Pennisetum purpureum*) or guinea grass (*Panicum maximum*) can be grown in this way, and the income from the stock will offset the loss of land for crops. Where land is communally-owned, or where livestock have free range after the crops are harvested, a grass such as vetiver (*Vetiveria zizanioides*), which is persistent but not palatable, is more appropriate.

The use of hedges which can fix nitrogen and provide mulch, such as *Leucaena spp.* and *Cassia siamea*, have been found very beneficial in certain situations. However, the drier the area the greater risk of competition with the crop for moisture.

Agroforestry
Agroforestry practices which have a role to play in certain circumstances include alley cropping using, for example, a row of *Leucaena sp.* or *Sesbania sp.* alternating with several rows of maize. Another agroforestry technique is planting trees and crops in combination, such as citrus or avocadoes with maize and beans. But the risk of competition of the woody species for light, nutrients, or water, must be kept in mind. The traditional system, found for example in Malawi, whereby crops are grown under a stand of *Acacia albida* that sheds its leaves during the growing season, appears to have considerably wider application than at present. The use of agroforestry species in conjunction with grass strips has much to commend it. The trees increase productivity, and the grass provides an effective barrier to runoff at ground level.

Structural measures

Structural measures have an important role to play in certain situations, such as farming steep slopes or retention of rainwater in dry areas. But they are only successful when farmers are convinced of the need, and install and maintain them voluntarily.

Cut-off drain
Cut-off drains are needed to protect cropland against runoff from a natural or artificial waterway or on to grass. They are normally placed above the cropped field. If, as sometimes happens, they are placed at intervals down a long slope which is cropped, they usually fill up with sediment and become ineffective.

Infiltration ditch
An infiltration ditch is basically a cut-off with zero gradient designed for the retention of runoff. They are usually made big enough to incorporate bananas and other fruit trees that can benefit from the stored water.

Terracing
There are many types of terraces, and they are given different names in different places. The simplest are sometimes referred to as bunds if there is a channel on the upper side, or they may be called narrow-based channel terraces.

Bunds and narrow-based channel terraces
Bunds may be formed from soil, stone, or a combination. They are often made on a gradient to discharge water. If they are made on the contour to retain water they must be bigger and cross-ties should be used to prevent lateral flow. The narrow-based channel terrace is made by digging a channel and placing the soil on the lower side.

Fanya juu terraces
These terraces are popular in certain parts of Kenya and are being tried elsewhere. They are made by digging a trench and throwing the soil uphill to form an embankment. They were originally seen as the first step towards the formation of bench terraces. The accumulation of soil above the embankment by the processes of erosion and cultivation assists in bench formation.

Bench terraces
Bench terraces may be made outward-sloping, level, or inward-sloping. Reducing land slope reduces the potential for erosion but where terraces are outward sloping, runoff is liable to pass from one terrace to the next. As a result rilling can occur. Outward-sloping benches often develop from *fanya juu* terraces and there is no sharp dividing line. Level bench terraces, if well made with a lip at the front edge, provide maximum control of soil and water losses. Inward-sloping bench terraces are also effective, but less common because the embankment has to be higher and more work is required in construction.

A Strategy for Better Land Husbandry at Thabana Morena

T.F. SHAXSON

Introduction

An integrated rural development project is under way at Thabana Morena in southern Lesotho. This paper briefly describes a proposed strategy for agricultural improvement, which aims to integrate higher production with increased environmental stability of an area whose resources of soils and range vegetation have, for many years, been seriously degraded due to poor management. Natural, economic and social factors interact in complex ways in rural life in Lesotho. They have significant effects on land use and agriculture, which for many families is not the main source of income. A large proportion of adult men are away earning money in the mines of South Africa.

The area rises from 1550m to 2100m amsl at latitude 29°45'S. Summers are warm to hot, but the winters are cold, with early and late frosts. The rainfall, mostly received over 21 weeks in summer between October and April, varies widely around an annual mean of 711mm. The project covers steep-sided basaltic uplands, predominantly under indigenous pasture. It also incorporates lowland areas, which are mostly cultivated, where soils are derived directly from sedimentary rocks and from downwash materials from higher areas. Almost all the land is of capability class III, or worse.

The human population is about 17 000 and rising at a rate of about 2.6 per cent annually. About sixty per cent of *de facto* heads of households are women.

About sixty-five per cent of the area is under forms of cultivation agriculture, with no room for expansion except on to very steep rocky slopes. Land holdings are between one and three hectares a family, often split into two to four parcels, and up to thirty to fifty per cent of the cultivated lands are left under sparse weed-fallow every year. Yields of the most common crops, maize, sorghum, beans, wheat and peas, are very poor. Little fertilizer is used because of great uncertainties about the weather during the cropping season. Large numbers of cattle, sheep and goats graze freely throughout the area, under traditional controls, during the year. Dung, paraffin and shrubs are preferred household fuels. Soils have become seriously eroded as a result both of geological and agricultural factors.

Problems

The main problems identified by farmers are:

- Drought and erratic weather conditions causing low yields
- Difficulties in getting inputs, due to lack of information about recommendations and to costs of obtaining requirements

- Some families with no source of cash income
- Insufficient draft power for ploughing and planting at optimum times
- Poor education of children and youths herding the animals, and
- People from outside agencies not listening to local people's views.

Other problems, discerned by the author were:

- Serious sheet and gully erosion, with continuing losses of topsoil from cultivated land
- Cultivated fields damaged by uncontrolled runoff from poor-condition rangelands above them
- Low investment in agriculture due to risks involved, and the availability of other sources of income
- Significant phosphorus deficiency in crops due to soil acidity in many parts
- Poor nutrition of animals due to scarcity of feed over long periods of the year
- Inadequate management of lands and of the animal/pasture complex
- Women not able to do the same work as men, and
- Outsiders' insufficient knowledge of the local environment and of attitudes and perceptions of the inhabitants.

Requirements

Continued application, over many years, of conventional approaches to raising yields and conserving soil have clearly failed adequately to achieve what was intended. The problems of the area are interwoven, and single-shot solutions would be unlikely to be effective. In arriving at adequate, appropriate, feasible and acceptable solutions, the ultimate decision-makers will be the farmers, households and communities themselves.

As a framework for future discussions and actions, a strategy was prepared for improving and sustaining the productivity of the land. The strategy had to satisfy the following requirements:

- Improvement in the security, quantity and variety of food produced
- Improvements in the chemical, physical and biological conditions of the soils both for plant production and for conservation of water and soil. These also provide opportunities for managed self-recuperation of degraded soils and range vegetation
- Concentration of the limited resources of rainwater, organic materials, chemical fertilizers, and other inputs to best advantage, so as to optimize plant production of crops, fodder and woody species
- Improvement of animal nutrition, for their better health and productivity, and for the production of manure
- Making it easier for women to combine their food-producing activities with other family-centred activities
- Creating physical and mental conditions in which households or communities themselves would allocate land for planting and managing trees and shrubs to satisfy a variety of needs

- Stimulating the processes of local discussion and decision-making about changes and improvements, and for avoiding a top-down prescriptive approach, and
- Progressing towards, rather than beginning with, discussions in the community about planning future uses of the land in the light of anticipated increased population pressure.

Land husbandry

The strategy was based on the concept of good land husbandry, which may be described as managing preferred systems of land use so that there will be no loss of the land's stability, productivity or usefulness for the chosen purpose.

Good husbandry is essential if land use is to satisfy present human needs without damaging its capacity to do so in the future, and without diminishing future options for alternative uses. Land husbandry is not just a substitute for soil conservation. Nor is it just a set of techniques, in the way that soil conservation is often understood. It connotes ways of thinking about and of managing the land. This includes knowing its characteristics, understanding its potentials and problems, predicting how it may react under proposed changes in its use and management, and manipulating its dynamic systems to maintain its health and productive capacity.

Because it is rural people that husband the land, good land husbandry is based on both social and ecologic principles.

Social principles

Rural people make rational decisions within the context of their circumstances, skills and resources. Improvements in land husbandry are more likely to occur if the context within which they make their decisions is changed and enlarged, than if attempts are made to change decisions within an unchanged situation. Landholders are more likely to be enthusiastic undertaking new activities if they have been full participants in the decision-making process that led to them, than if they have been inadequately consulted.

The social principles are:

- To help people do better what they already do
- Allow people to rank their own problems
- Recommendations are more likely to be adopted if they increase output
- Many problems are interlinked. Solutions should be in the context of an ecology of disciplines, not unrelated bits of information
- Technical advisers should act as catalysts and sources of new information to be discussed
- Conservation is more likely to be achieved by stealth, by integration of conservation practices into the production system, and
- Improvements in land husbandry practices should be developed from the knowledge and priorities of farmers.

Ecological principles

The principles are:

- Sustainability requires achievement of a biomass equilibrium, so that removal of biomass from a given place is balanced over time by its renewal through biological processes
- Accelerated erosion is a foreseeable ecological consequence, not a prime cause of land damage
- Good land management achieves conservation. In this regard, adequate amounts of cover to the soil and a sufficient grade and stability of soil structure are of primary importance
- Organic matter and organic processes are particularly important in ameliorating soil conditions, especially in terms of promoting good soil structure and aeration, the retention of nutrients, and retention of plant-available water
- Land husbandry methods should be adjusted to minimize degradation in an area. This can be done by use allocations that avoid a part wearing out more quickly than another
- Specialized biological measures and engineering structures should be used only to supplement good land husbandry, not to replace it
- Sustainable agricultural productivity may require regular periods of restorative treatment within the cycles of agricultural use
- Increased production of plant parts for each unit area may automatically provide increased protective cover to the soil, and thus contribute significantly to conservation
- The more diverse and complex an agricultural system, the more stable and sustainable it will be
- As much land as possible should be maintained under flexible types of use, thereby leaving open the greatest number of options for different uses from which others may choose in the future
- To a feasible extent, multiple uses of the same area of land should be favoured, so as to maintain as much as possible elsewhere under undisturbed conditions, and
- Descriptions of suitability of land for different uses should always connote definitions of both the erosion hazards of the land and the conservation-effectiveness of the proposed forms of use and husbandry, so that they can be balanced.

Strategy

The strategy which was eventually derived proposes a sequence of activities as guidelines for the extension agents when they participate in the communities' discussions. Each stage would follow the other at appropriate times, as and when families and communities considered most appropriate.

Stage one – improve the variety and security of nutrition of village households by developing, intensifying and utilizing the productive potentials of households' round-the-house (RTH) plots. This is based on helping women in particular, by limiting distances travelled to collect water and fuel and helping them to improve gardening skills. It would emphasize the concentration of inputs and efforts on to limited and intensively-managed areas.

Stage two – develop fodder sources in open fields, as well as in RTH plots, for families' own uses, as well as for barter or sale, once greater food security had been achieved in stage one. Soil regeneration in outer fields would be more rapid and effective under semi-perennial fodder resources than under continual low-output cropping. Animals would improve as a result of easier availability of fodder. Drastic specialized actions would be needed to improve the severely-degraded rangelands.

Stage three – develop resources of woody plants for the variety of purposes decided by individuals and communities. Allocation of land for such use would be more readily considered than at present, once people felt that supplies of food and fodder were reasonably secure after passing through stages one and two.

Stage four – progressive adjustment of overall land use patterns. This should conform, as closely as desirable, with the types of use required by the communities, and with the land characteristics of the area as a whole. Land use patterns must adhere to communities' discussions about the hazards of degradation of the land's productive potentials, and to future demands on the land imposed by future population increase.

Conclusion

An unconventional way of looking at land degradation, which is often found in developing countries, and attention to social and ecologic principles of land husbandry, has led to the definition of a strategy which indicates how an ecology of disciplines could help to solve an ecology of problems. It was done through a sequence of actions suggested for discussion by the rural families of Thabana Morena.

Smallholder Adoption of Some Land Husbandry Practices in Kenya

R.J. CHEATLE and S.N.J. NJOROGE

Introduction

The Kenya Institute of Organic Farming (KIOF) has made good progress since it started in 1986. It now has 42 self-help groups in the Central Highlands of Kenya. Group membership ranges from 20 to 40 people, some are church related, others are lay groups. KIOF employs a small number of promoters, to help initiate self-help groups. Some live within one of the communities served. Promoters teach organic farming practices to groups, which are followed up. They also encourage individual smallholders. New groups are formed, but only after a request has been received to do so. A programme of exchange visits provides a basis for farmer to farmer networking.

Approximately forty per cent of those who commenced organic farming with KIOF fell away at a relatively early stage. But about 2800 smallholder producers have adopted land husbandry practices of composting, double digging, and water harvesting with enthusiasm over the past three years. These techniques are labour intensive, and some professionals argue that labour intensive practices are not readily adoptable. That any technology has been adopted is a good reason for an investigation.

The authors visited groups over a period of 12 months. Some groups had three years of experience (six seasons), while others had two years. Information was obtained from group discussions with individual farmers, and in-depth discussions with key informants. In this paper information is reported to describe what some smallholders are doing. Smallholders adopt practices because they perceive benefit or advantage. Pooling the individual cases enables us to make some general statements. The paper also shows the need for applied research to accompany innovative implementation work. The purpose should be to identify information that will enable improvement of smallholder farming systems, and also the dissemination of that information.

The Central Highlands region

The region is one of physical and cultural diversity. There are old upland massifs around Machakos, but much of the area is tertiary volcanic country. Altitude varies, but 1200m is typical. Temperatures range from 4°C to 33°C. The bimodal rainfall regime ranges between 800mm and 1300mm, with considerable variation in the seasonal rainfall totals. Periods of water stress within seasons are common, and dry spells between seasons always occur. Water supply is a major determinant of crop production.

The loams and red clay soils derive from the volcanic rocks and have been described as rich. Within the naturally moist woodlands the organic matter content of the soils is associated with a high capacity to absorb nutrients, good water-holding capacity, and good physical conditions in the upper soil layers. Most of those moist woodlands have been cleared, and smallholder land use is dominant, with small farm units of 0.5 to 3ha, though larger units do occur. There is always some subsistence production and usually part of the land is given over to coffee, tea or some other cash crop. Maize and beans are important subsistence crops, but other food plants are grown. Many farms often have trees to meet firewood and timber needs.

There is considerable land pressure. Population density is approximately 200 people per km^2, while farm size is diminishing through inheritance splits. Landowners are usually prepared to take action if degradation endangers production. As this country is deeply incised, homes tend to be located on narrow ridges, and water supply is far below in the valley.

Gardens are invariably on moderate to steep slopes, between home and water, with a soil erosion hazard.

Some eighty-five per cent of the people live in the highland country of Kenya. There is movement from the highlands to the semi-arid lands because making a living in the highlands is difficult. The government recognizes the need to contain this movement and to promote practices to enable sustainable livelihoods in both highlands and semi-arid lands. Less attention is given to the notion of rehabilitation and intensification of production in the highlands, although better rainfall there offers more opportunity for improving sustainable production than in the semi-arid lands.

Population in the highlands has grown rapidly since independence. Woodland clearance and use of more land on steep slopes has increased soil erosion. This problem has been tackled through a major soil conservation programme within the Ministry of Agriculture. Terraces, cut-off drains, and drainage ways have been emphasized. Now the need is to identify more acceptable, less costly, and less labour-intensive conservation measures. The potential of agroforestry is being explored.

There is less awareness of soil fertility decline. Once lands are brought into agricultural production the organic matter content in the upper layers rapidly declines with adverse effect on chemical fertility and physical properties. Growing two crops in the two annual seasons will decrease fertility if nutrients are removed in the crop and crop residues. Soil erosion differentially sorts and removes organic materials and finer particles, contributing to the problem of fertility decline. Smallholder producers often refer to a history of declining yields. Measurements of pH suggest that this may be due to increased soil acidity at some higher elevations. Observation of nitrogen-deficient maize of poor status in reasonable rainfall conditions adds another factor. The decline in organic matter will also be associated with inferior soil structure, increased runoff, lower infiltration, and less available soil water for crop growth.

None of the current farming systems includes practices to maintain soil fertility. On the small farm units it is difficult to use traditional approaches, such as fallowing, when scratching a living from a small parcel of land. Any soil fertility that is removed or destroyed is not put back or restored, though smallholders attempt to address poor soil fertility by using animal manure, which is a valued commodity. Recommendations from national research institutions invariably include advice to use inorganic fertilizer to sustain yield. This advice can only be partially followed, at best, because smallholders cannot afford the high price of fertilizer.

The limited empirical data available suggests that farming systems in this region are caught in a spiral of fertility decline. There is urgent need to explore methods for rehabilitating and conserving the fertility of lands. Since fertilizer is no panacea, all promising forms of organic matter management should be explored. Smallholders will adopt those practices which they perceive as beneficial.

The fundamental technical constraints are the need to:

- Improve water management and the efficiency of water use
- Rehabilitate or sustain soil fertility, and
- Minimize accelerated soil erosion.

The KIOF approach to composting, double digging, and water harvesting

Suggested methods for composting and double digging are described in KIOF (1990). Making standard compost, as taught by KIOF, converts crop residues, weeds, household organic waste, hedge cuttings, grasses, and the like, into a partially mineralized material. Relatively small proportions of animal manure (if available) and topsoil are layered into the green materials. These materials are turned and mixed twice every three weeks, and the compost is considered ready for use after nine weeks.

Smallholders with cattle or goats should site the composting pits near to and below the animal pens, to simplify transfer of animal manure into the heaps of material. Large heaps of compost are frequently made, and may be 1m high, 1 to 2m wide, and several metres long. Some smallholders make up to 24 tonnes of compost annually. While guidelines for use of compost are given by KIOF, the smallholder is free to use his compost at will, and no specific directions are given about timing of application, application rates, and placement. The farmer is free to experiment and develop his or her own methods to suit individual conditions.

Application of compost is likely to bring benefits. Organic materials within the soil system may adsorb water up to ninety per cent of their own weight, and have low plasticity and encourage granulation in soils, which promotes porosity. Cation adsorption is increased since organic colloids have two to 10 times the adsorption capacity of mineral colloids. In short, the application of compost will improve soil fertility. Sanchez (1976) provides a useful review of the beneficial effects of organic matter in soils, and applications of animal manure.

Double digging is recommended for intensive kitchen gardening close to the house or near a source of water. The practice combines the use of compost with a deeply dug bed. An upper layer of soil is removed, and a second layer dug over before soil mixed with compost is returned. It is a laborious process to set up double dug beds. Benefits arise from composting and from improvements in the physical condition of the soil. Deep digging will break up hard layers. This improves porosity and helps root growth. Nicou and Charreau (1988) have demonstrated strong positive relationships between improved porosity, root growth, and crop yield in sandy soils. Cheatle (1991) has drawn attention to constraints to tree growth of compacted soil layers in clay soils. Trouse (1979) has provided a useful summary of the relationships.

Water harvesting is storing the runoff from house roofs. Near the house a large basket is woven with pliable stems over a prepared foundation. This basket is waterproofed by several layers of cement. A pipe and tap is set in at the lower end. Guttering and pipework lead water into the tank. About 10 bags of cement are needed to complete each tank. Construction costs are low and the tanks are built by group activity on a self-help basis.

Completed tanks are similar to traditional grain storage facilities and may have a capacity above 1000 litres. The stored water is used for domestic purposes, for animal consumption, and to water double-dug beds.

Composting, double digging, and water harvesting in tanks may all be classed as practices to assist conservation. In each case there is a foundation from which other activities can be developed. Actual practice is amazingly diverse. Smallholders experiment a good deal and derive related techniques for themselves. The practices are not a package thrust at farmers, but can be manipulated and fitted by smallholders to meet their own perceived needs. This fact may be critical in explanations of adoption of these practices. KIOF promoters present the opportunities and subsequent discussions with smallholders may identify possible applications. This approach is quite different from the Training and Visit system, where a bit of technology is thrust at smallholders with little attempt to involve farmers and respond to their perceived needs.

Profiles of four smallholders

Four profiles report smallholder experience with one or more of the three practices described. Individuals and groups have quite different priorities, so the response varies. However, there are common features, especially in relation to the practices used and explanations of the benefit derived. In these profiles we do not aim to describe all the organic practices and conservation measures used. Instead we highlight certain activities and describe the rationale.

Peter Ngugi

Mr and Mrs Ngugi have a smallholding in the coffee zone of Muranga District. With help from neighbours, Peter Ngugi constructed a water tank close to his house. He and his wife have helped others to build tanks, and today several families have them. Near to the water tank are double-dug beds, which supply a range of vegetables through mixed and relay cropping. Cabbages, onions, carrots, beans, sweet and ware potatoes, aubergines, and tomatoes have been grown in a single bed. Several macadamia and citrus trees have been planted near, but downslope, of the beds.

Two cows provide a cash income from their milk, and manure for compost. Compost is regularly supplied to the double-dug beds. There is about 1ha of coffee and 0.5ha devoted to maize with beans, other subsistence crops, and napier grass for fodder. Like most farmers in the region, Peter complains about poor coffee prices and late payments. He is disatisfied with coffee as a cash crop, and would willingly shift into something else. In contrast, Peter is very supportive of organic farming and the family water tank.

This holding is within the high and medium potential zone of Kenya, and farming depends upon the rainfall. For Peter and other smallholders, water is seen as the main on-farm constraint. However, when KIOF staff first arrived it was the village women who immediately took up the question of water, because they have to fetch all the water from the valley far below. With both men and women in accord at an early stage, co-operation to promote a community water tank project was readily agreed. This was associated with double-dug beds. KIOF provided technical support and a little seed money for cement to get things moving. A promoter visited regularly to encourage and advise.

Mr and Mrs Ngugi recognize that the water tank has changed their lives. Water is now available most of the time for domestic use and for the vegetable beds. Mrs Ngugi and her daughters no longer spend long periods in the laborious task of water collection and the new supply is potable. Instead of eating mainly maize meal and beans the family now has a wide range of vegetables available over most of the year. The Ngugi family also perceives that a better food supply has improved their health. Money is always a vital consideration when one has very little cash in pocket. Prior to the double-dug beds Mrs Ngugi frequently sought money from her husband to buy vegetables for the family. Sometimes the requests for money led to domestic conflict. The need for money to buy vegetables has largely fallen away so now Mrs Ngugi is sometimes able to obtain a little money for herself from small sales. These are benefits of intensive land use around a water tank.

A benefit not recorded by smallholders is that this intensive form of land use can lead to less pressure of use on other lands. It may mean that good conservation measures can be practised as part of vegetable production, and that less suitable land on steeper slopes can be used for longer-term crops that protect the land better, such as napier grass. Around homesteads, land is often bare or has little cover. These areas are gathering grounds for runoff. Double-dug beds are set across the slope and act as a barrier to runoff and to soak up surplus water. They are an important conservation measure when sited appropriately.

Mrs Christine Karuru
This lady is a commercial farmer in the coffee zone of Kiambu District. The family is relatively well off. Mr Karuru owns a minibus. Profits from it and the farm business have enabled the family to buy three farm units, each about 2ha. Christine is enterprising and a committed Christian. As well as conducting her farm business she spends time helping others to develop organic farming. Her parish priest knew of KIOF and approached the organization. Christine and others were helped by KIOF to attend a short course in organic farming at Ngong Training Centre. This experience convinced her to try organic farming and to give up some of her time to help others.

One of the farm units with steep slopes and red clay soils was dedicated to coffee. The trees are established on bench terraces in accord with Coffee Board recommendations, which also include the use of grass as an anti-erosion measure, no other crops to be planted in the coffee, and the use of fertilizer.

Fertilizer was previously a considerable outlay, but Mrs Karuru no longer uses fertilizer and regards this as a saving and a great benefit. She argues that her coffee yield has improved over three years of composting. This may be true, for the compost is well dug in, whereas it seems likely that the fertilizer was not. In which case, much of it may have been washed off the surface. Time and rate of application may also have been less than optimum. Unless reasonably efficient use of fertilizer can be assured, its recommended use is questionable. This is because the cost of fertilizer is high and some of it may be wasted.

In transferring to organic farming, a first task was to remove what were considered as 'unmanageable and useless grasses' from the coffee and interplant with maize and beans, using compost instead of fertilizer. In the first season, accelerated soil erosion was noted. The response to it was quick. Mrs Karuru built a cut-off drain, a spillway and three *fanya juu* terraces to protect the land. Part of the rationale for siting the spillway along a boundary was that it would protect her land from grass invading from the neighbour's farm. Bananas and fruit trees have been planted into the contour ditches, so no land has been lost to conservation. The holding is protected and Mrs Karuru is very pleased with her crop diversification and increased production from the land. She argues that coffee yield has increased and that she has the benefit from the additional crops that are also yielding well. Composting and improved soil moisture due to a decrease in runoff may have improved yields. This mixed planting approach has markedly increased cover and hence soil protection during the main part of each rainy season.

Large quantities of compost are made. Manure from four cows is mixed with grass, banana leaves, and coffee pulp collected from a nearby coffee processing plant. Boundary hedges are also clipped to provide material. At present, Mrs Karuru is applying compost every two years. She is experimenting with frequency and manner of application. Her present method is to insert compost into a contour furrow, and plant on, or close to, that furrow. Observations of maize growth suggest that a further application at present rates is required after two years (four seasons). KIOF has introduced amaranthus varieties and some farmers have identified poor growth of this plant as a useful indicator of when fresh applications of compost are required.

Peter Njenga

Like many other smallholders Peter Njenga is an experimenter. His holding is about 2ha in Kiambu District. He has taken to organic farming and, by trial and error, worked out improved farming methods. One consists of a row of deeply-dug holes into which compost is worked. Maize is densely planted round the edge of each hole, and later thinned to about a dozen plants. The ground between these maize-pits is planted to other crops, mainly tomatoes or beans. The next season, new holes are dug between the existing ones, and these are used for composting and planting maize. In the third season this area is made into a raised bed and kept that way for two seasons. Two rows of sweet potatoes and two rows of beans will be grown in each season. Therefore there is a rotation system between beds. There are several grevillea trees on the land.

The raised contour beds are an effective soil conservation barrier. Complete cover of sweet potato foliage is likely to be established within 30 days. This protects the soil from erosive rainfall. The mixed cropping also provides good ground cover. Mr Njenga knows that his practice protects soil. He is also pleased with the effects of composting after some seasons. He argues that his crop yields have improved, and that he is getting more crop off less land. He is conscious of the fact that intensive production on a site close to the homestead has time-saving advantages. The grevillea trees are considered useful as a timber source and also because 'crops are better' near them.

John Kagwe

John Kagwe inherited 1.5ha of sloping land as his share of an 8ha farm unit in Nyeri District. That small parcel of land supports nine people today by organic farming. This was not always the case. About one third of the land is under tea. Until 1987 this was the main cash crop. The gross annual income from tea for 1987 was about 11 000/-, and there was subsistence income in the form of maize and other crops consumed on the farm. There was no stock. Foodcrop yields were considered low, and food supply for the family was inadequate. John Kagwe had been granted a loan of 7000/- in 1986 to buy fertilizer and chemicals. Servicing the loan, paying medical bills and sending a child to school used all remaining cash and there was nothing to spare. By 1987 John regarded himself as near bankrupt and believed he was in danger of losing his land. His wife ran away.

In desperation John Kagwe was prepared to try anything. He moved into organic farming when he was selected for a short course at a training school for organic farmers. When he returned from the course he applied many of the ideas that he had gleaned.

Many of John's decisions were economic. Tea was not producing sufficient income. Smallholders appear to receive a lower price than commercial farmers. John was concerned about the labour inputs to tea. Picking was done largely by his wife and a daughter, though he had to help and occasionally a picker would be hired. The walk to the weighing station was about 2km. Often there was a long wait for the tea lorry, sometimes until midnight and beyond. Tea production was taking up too much time for too little money.

After his wife left, he decided to rent out the tea bushes for the year at 2/- each, to bring in about 6000/-. Using some of this income John was able to pay off the balance of his loan, and to buy a few poultry, rabbits, and pigeons, and to set himself up as an organic farmer. The stock helped to bring in a little income until other enterprises were underway.

With more time available John worked at his organic approaches. Composting was one of the first, and is supplemented by the preparation of liquid manures. Throughout his period on the farm he maintained a small stand of *Cupressus* and *Grevillea* to one side of his tea. In addition he has always kept a few *Commiphora zimmermannii* for yam planting. He extended his use of this tree for planting passion fruit, and planted the trees along the contour where tree roots will help bind the soil. *Fanja juu* terraces have been constructed and raised beds also protect the land. Rows of trees and perennial plants are now interspersed between field crops. John has also constructed water harvesting pits and has an ingenious method to supplement water supply to double-dug beds near his house. Between two large crescent-shaped double-dug beds is a hollow into which twigs and other residues have been placed. Furrows lead rainfall into this area from which it can seep into the beds. He has extended his double digging to quite large beds for vegetables and fruits. He has also constructed some pits for water harvesting. In discussion John Kagwe does not dwell large upon soil and water conservation, though he does recognize that what he is doing helps his plants to grow and keeps his land in good shape. He has successfully integrated water management, maintenance of soil fertility, and soil erosion control into his farming system.

John Kagwe's assessment of his organic farming is 'my wife is back now, there is increased yield, we have more money, we have been able to send the children to school and we have more food and are healthier'. Only 420/– was spent on medical bills in 1990, but over 4000/– was spent on school fees, building funds, books, lunches, and uniforms in order to put five children into primary school grades. Clearly there have been changes for the better as a result of the changes in on-farm enterprises. Income now arises from several new sources. There is rental income from tea, and income from sales of rabbits, chickens, passion fruit, tree tomatoes, tomatoes, and seedlings. The fruits are sold at the farmgate and at the market in Karatina. A minimum estimate of income for 1990 is:

Tea rent	6 000
Fruit sales	12 000
Animal sales	5 000
Total	23 000

The main expenditure has been 1800/– on bought-in chicken food. No fertilizers or chemicals have been bought. During the year John had sufficient cash in hand to purchase a cow at 13 000/–.

Summary of the profiles

The cases reported demonstrate several important aspects of on-farm development of good land husbandry practices by smallholders themselves, and the adoption of conservation practices. All this arises from the initial teaching of three basic techniques, composting, double-digging, and water harvesting in tanks. The smallholders see that their action to adopt and develop particular conservation practices is beneficial, and that a wide range of benefits is the result. Good land husbandry or conservation has been successfully built into farming systems to improve water management and make more efficient use of water, improve soil fertility, and protect soils against erosive rainfall.

As well as conservation there appears to be improved production and more varied produce. An important result is the additional revenue arising from more sales. Purchase of fertilizer and other inputs appears to be a constraint upon development, particularly with low farmgate prices for produce. Organic farming removes this constraint and allows other opportunities to be taken.

Within three years about 2800 smallholders have adopted conservation practices, and made advances on their farms in respect to both conservation and production. This has been achieved at low cost and with very limited manpower resources. It is useful to emphasize that one or two well-motivated people can get things moving. Identifying such people is not always easy, but once they are known their efforts should be supported to enable appropriate forms of consolidation and expansion. The flexibility of the NGO, and the speed with which linkages and implementation activities can be commenced, also suggests this kind of operation deserves support.

Technical requirements

In 1975 the FAO produced a bulletin called *Organic Materials as Fertilizers*. The state of knowledge was reviewed and attention was drawn to large knowledge gaps. Thirty-eight major research recommendations were listed, only three of which were about social and economic implications. In spite of the bold initiative very little has been accomplished since.

Given our general understanding about the importance of organic matter recycling, and the need for good soil physical conditions for plant growth, it is surprising that so little scientific information is available for application at the producer level. Throughout the African continent, inorganic fertilizer trials have been a dominant theme for many years. Much less work has been done on organic materials, animal manure, and green manures. The experiments have generally been related to field crops, for example manure applications have been spread and tilled into the topsoil. The results show that organic additions are beneficial, but that relatively large applications are required. Experiments have shown the value of deep tillage. The information suggests that composting and double digging will improve crop growth. Experimentation with composting has been limited, but one 10-year experiment confirmed the value of the practice (Rohde, 1954).

Little information is available to KIOF about such important matters as the most efficient mix, and methods for making compost, application rates, timing of application, manner of application, and benefits of surface mulching versus digging-in. Many questions need to be resolved about composting, double digging, and water harvesting.

Smallholder producers have a role to play in the process of knowledge advancement. Those working in crop production experiment frequently. Sometimes they are ahead of more formal research processes. In this paper are two forms of application of compost close to the crop, an application in furrows and one in holes. The standard literature does not cover this approach well. The smallholder experimenters have inferred that concentration is beneficial to the crop. Two of the people profiled have developed rotations both for composting and for crops. The developments are interesting since Sanchez and Cochrane (1980), and others, have drawn attention to many physical and chemical soil constraints in farming systems. These include shallow depth, water stress, low water-holding capacity, erosion hazard, soil compaction, steep slopes, poor nutrient status, and soil acidity.

The potential of smallholders as researchers is demonstrated since they can contribute to applied research. The opportunity to be grasped is that of forging the linkages and using the trained researcher to motivate and guide smallholder experimentation. The role would include enhancing the scientific validity of experiments and assuring wide dissemination of the results. This also suggests another need, a review of formal research. NGOs, like KIOF, need, but do not get, research support from the national research institutions. It would be an effective use of available manpower if a few research workers could be assigned to work with NGOs and establish links with them and extension workers. Two of the innovative practices described here need further investigation, and dissemination if their usefulness is confirmed.

Social and economic requirements

Increased farm income is seen as the most important benefit by the smallholders participating. The second interest is the best use of available labour. They would like to know something about the opportunity cost of one activity versus another, perhaps using simple models. There is need for information about the economics of sustainability in farming systems. For example, there is little information about reduction of crop yield, and the cost to smallholder and nation of accelerated erosion.

Markets and marketing are also a neglected area of smallholder microeconomics. For smallholders there is no point in production if you cannot sell the product at a profit. We need applied research close to the grassroots about smallholder production, market-led demand, marketing opportunities, and how to take advantage of short-term and long-term opportunities. This is relevant to all conservation or sustainable practices. Without financial incentives the practices will not be implemented.

The KIOF operation is simple, cheap, and effective. More information is needed about just what is being achieved and the costs and benefits. This is relevant in relation to donor-funded development. If KIOF has an effective approach we need to know how it works, and to spread that message. It is high time that some role model projects were developed and scrutinized by high-profile monitoring and evaluation to provide the key lessons that others can apply.

The smallholder profiles tell us something about decision-making, and we need more information about this process. Adoptions of revised or new practices are critical to the development of sustainable farming systems, and we need to know more about how individuals' and groups' choices are influenced.

KIOF has spread some messages effectively, because adoptions have occurred. This provides an opportunity to investigate dissemination and adoption processes to identify just what the successes are and how they have occurred. Anthropological studies are required to gain a better understanding. This is necessary for the planning of additional activities. The KIOF operation has included some smallholder to smallholder day exchanges. These have been effective in that some smallholders have seen organic farming practices for themselves, interacted with others, and then done something positive after the visit. This limited experience suggests that farmer-to-farmer networks should be developed through the NGO system, with linkages for a two-way exchange of information with national research and extension systems.

These examples of the KIOF approach are small, and each affects only a handful of people. But they are important because, as Paul Harrison said in *The Greening of Africa*, 'Our success stories are like seeds. If they are widely enough sown, they can take over the field.'

Cash Incomes and Conservation: increasing both simultaneously

S. CAIGER

Introduction

Farmers' fundamental priority is to increase their disposable incomes in the short term. They will do this even if their agricultural and management practices are likely to cause a decline in income later. This is seen as entirely logical: the poor do not have the luxury of resources to invest in the short term against the probability of higher returns later. Therefore, we assume that if conservation management for soil and water resources is to be adopted by producers in marginal environments, it must be a consequence of either cash crops introduced into the farm, or new practices that increase incomes from existing cash crops.

In addition to income, producers will assess returns to labour. No one will take on a massive increase in workload for a small increase in cash returns.

The early work in Africa on the benefits of manure, fallows and rotations gave a clear indication of the farmers' priority for this season's income. The volume of case studies and experiments done since have not altered the basic premises, and do not need further review here.

Background to crop selection

This paper identifies some products with market potential, from crops which are suitable for small farm operations and which also improve soil and water management.

The selection process is market-driven. Successful ventures give the market what it wants, when it wants. The market is the point of entry to crop selection, and agronomists, soil scientists, and the others, cannot stand back from it. They must be familiar with the current status and future trends.

Crop selection requires multiple input from agronomists and soil scientists to identify suitable crops with beneficial management practices and effects. Sociologists identify crops that can fit in existing farm systems. Economists assess markets and profitability to farmers. Conservationists identify sustainable production systems. Engineers assess the processing requirements and options. Business managers evaluate processing enterprises, while financial institutions assess financing requirements. On the basis of this resource audit, sensible crop selection can be made from the market demand. Building financial models for crops and farmers, showing their labour and input requirements, and building and understanding profit and loss and balance sheets, are fundamental requirements.

The essential oil crops

The essential oil, or aromatic crops, yield an essential oil when steam is passed through the cut crop. The steam dissolves the oils, and when the steam is condensed, the oil and water separate and the oil is drawn off. The oils are concentrated in different parts of the plant in different crops: in leaves and stems in the grasses, such as citronella, lemon grass, palmarosa; in leaves in geranium, eucalyptus; in flowers in chamomile; in seeds in anise; and roots in vetiver.

For all essential oil crops, oil yield is dependent on herbage yield and oil content of the herbage. The latter is dependent on variety, environment, and agronomic practices.

Sourcing of material with high oil content is important, but climate has an active effect, and optimum agronomic packages must be developed – with oil yield and quality the determining factors. The effect of fertilizers on oil quality and yield must be determined. In many crops, oil quality can be adversely affected by fertilizer application. Oil content of the cut material is further affected by harvest interval. For different crops, in different environments, the optimum frequency of cutting must be assessed to maximize oil yield.

Before considering a few of these crops in more detail, it is worth outlining their general advantages:

- They are all high value, from citronella at the bottom at around US$4 a kg, to chamomile at over US$800 a kg
- All the products have specifically defined chemical quality. High quality products will always find an outlet in this market

- These are the natural flavours and aroma chemicals, and with the trend to natural products in the west and elsewhere, their market share is expanding at the expense of industrial, artificial chemicals. Both market demand and future potential are strong
- Many are used as cheap scents in soaps, so there is often a starter market within potential producer countries
- High value also results in minimum commercial shipment costs, since quantities are small – a drum or two can be sold, there is no need to fill a container. This is particularly important when starting a commercial enterprise
- High value also eases the transport problem. Air freight is practical
- Products have a long shelf-life, so can be stored for a long time prior to marketing if required. Lengthy transport, or bulking up small quantities to a commercial shipment, is not a problem
- The product is oil from the distillation. All the vegetative matter is left as spent material in the still, and can be returned as mulch to the crop
- As high value crops, they can give significant cash income from a small area. Therefore, they can fit in to areas where there is acute land shortage, and can be considered as strip crops in existing crops, and
- Distillation requires processing businesses, but not necessarily a high technology process. Processing can be developed as a growers' co-operative venture, or as a separate local enterprise. Distillation plant can easily be mounted on a small lorry, or trailer. Thus it can be used to service an area comprising many villages, if cultivation is in small widely dispersed areas.

Citronella, palmarosa, and lemon grass
This group of grasses covers a wide environmental range, but all share similar agronomic and management characteristics:

- The crops are all perennial grasses, and can be grown for a long time (five years or more) before replanting is necessary. They are ideal as erosion and terrace barriers
- Oil distilled from the cut leaf and stem material. Leaves can be cut by hand or machine. They are then left to wilt after cutting to reduce bulk. Spent leaf material can be returned as mulch
- Spread of grasses as weeds is not a serious problem. Grasses either do not flower, or are harvested at flowering before seeds are mature, and
- There is always the possibility of developing a small market for the dried leaves as teas.

Lemon grass (*Cymbopogon spp.*)   A range of species are referred to as lemon grass – *C. citratus* (West Indian lemon grass), *C. flexuosus* (east Indian, Cochin or Malabar lemon grass), *C. pendulus* (lemon grass). The oil is used in soaps and perfumery for low-cost fragrances. Citral can be used to give a synthetic perfume with the scent of violets, and is used in the synthesis of vitamin A.

The species require warm and moist conditions for good growth, though lower production can be obtained from more marginal environments. Rainfall levels are ideally less than 2000mm, but this can be less particularly where humidities are high. Drainage should be good. Altitude is not a problem, but temperature fall will reduce growth and production. Plants form clumps, and can be grown in rows, or planted on the square (60×60 to 90×90cm is a typical range).

Citronella grass (Cymbopogon nardus and *C. winteranius)* Two types of citronella oil are traded – Ceylon citronella (*C. nardus*), and Java citronella (*C. winteranius*). Ceylon citronella is sourced almost entirely from Sri Lanka. Java citronella is grown widely in Asia, Central and South America. Both types are used widely in perfumery, mainly in the cheaper range of household and industrial products. Citronella can be isolated and is used in building many perfumery products. The growing conditions are the same as for lemon grass. Propagation is by root division. Plant spacings range from 0.5 to 1.0m on the square depending on fertility.

Palmarosa grass (C. martinii var. martinii) The grass occurs in the drier parts of India, and is cultivated in Indonesia, Malaysia, and other countries. Palmarosa oil is used in perfumery.

Propagation is by seed, and a fine seedbed should be prepared, the seed sown, and the site rolled, or raked. Seed rates of 10kg/ha can be used, or nurseries established. A very wide range of sites could be tried in Africa. Optimum rainfall levels are in the range 800 to 1000mm, but lower levels will support the crop with lower yields. There is no altitude restriction. The crop is a heavy feeder, and high yields require additional nutrient application. The crop is cut and distilled just after flowering starts. Three to five cuts a year are taken, depending on growth and conditions.

Ginger grass (C. martinii var. sofia) Ginger grass grows in wetter, more poorly drained soil and lower altitudes than Palmarosa grass. The oil yield is higher, but is of lower quality. Yields are in the range 90 to 115kg/ha.

Vetiver (Vetiveria zizanioides) Vetiver is a densely tufted perennial grass. It is native to India, Burma, and Sri Lanka, but is now widely grown, frequently for erosion control. The crop is cultivated in Java, India, Reunion, and Central and South America. Zaire and Angola have a history of exporting. The roots yield an essential oil which is used in perfumery (soaps and cosmetics), and as a fixative for more volatile oils. The roots are also used for making mats and fans. Young leaves provide good grazing. The whole plant has to be dug up at harvest, so in fragile environments, harvesting should only be in strips among sequential plantings. The spent material can be returned as mulch so that erosion dangers on harvested areas are minimized.

Some varieties flower, and can be propagated by seed. In others, propagation is by root division. Planting is often close, to maximize the density of roots at harvest. A typical spacing would be 20×100cm. Plants grow in large clumps, with stems arising from a much branched rhizome.

As the roots are harvested, light, sandy soils are best, as roots are easily lifted and cleaned. Rainfall levels of 1000 to 2000mm/yr are optimum for production, but the grass is tolerant of lower levels of rainfall. It is grown in the tropics and sub-tropics, and at altitudes of up to 1500m. Regular cutting of the grass increases root developments, providing further mulch or fodder.

At harvest (during the dry season) the grass is first cut and removed, or burnt. The roots are then lifted with a fork, washed, and dried in the shade. Yields of around 1000kg/ha dry root can be obtained.

Distillation is best done using live steam generated in a separate steam boiler. The roots are chopped finely just before distillation. Distillation takes a long time (18 hours or more) and requires careful control. A small commercial distillery would need about 40 to 50ha of vetiver to keep operating during a six month dry season, and given the 15 to 24 month growth period before harvesting, about 80 to 100ha should be under cultivation at any time.

German chamomile (Matricaria chamomilla)
Chamomile is also classed as a medicinal plant. The flowers are picked, and can either be dried for use in herbal teas, or distilled, yielding a very valuable oil. The crop is a short-term annual, and well suited for cultivation in small plots on family farms. Simple hand-held combs can be used to harvest the flowers which give acceptable rates of picking for family labour. The major producers are eastern and western European countries, Egypt, and Argentina, but many other countries are assessing the feasibility of production. The main characteristics of the crop are:

- The harvest portion, that is the flowers, is very small and when harvesting has finished the crop is still green, and can be dug in as a green manure crop
- The crop gives total ground cover soon after germination, so erosion control is excellent, and
- The young plants are frost hardy, so are suitable for highland areas.

The crop is propagated by seed. Seed rates are in the range 2 to 4kg/ha. The optimum environment has a cool period to the start of the growing season, to slow early growth, with the main growing season having a good mixture of sunshine and rain.

Extended excessively high temperatures (more than 35°C) during the growing season will shorten the productive life of the crop. The crop is tolerant to saline soils. The crop responds well to organic manures and composts, rather than artificial fertilizers. Great care must be taken with the interaction of artificial fertilizers and oil quality.

Eucalyptus (Eucalyptus spp)
Eucalyptus oil is the worlds most traded essential oil. Two different types of oil are produced – perfumery oil, from *E. citriodra*, and medicinal oils from several species. China and Brazil are major suppliers of the perfumery oil. China, Brazil, southern Africa, Portugal, and Spain are major

producers of the medicinal oil. All species are coppiced once or twice a year, depending on growing conditions. Planting can be done on a small or large scale.

The main characteristics of interest are:

- Many species are adapted to very marginal, dry, environments: *E. polybractea* has a natural rainfall range of 350 to 600mm/yr
- They are permanent tree crops, with clear advantages for soil and water management
- Only the leaves are distilled. The twigs and branches are available for firewood. Spent leaves can be returned as mulch, and
- Most species have timber uses, and oil production can be combined with timber production to provide early returns to a longer term crop.

All species are propagated by seed. Seed is exceptionally fine, and a few hundred grammes of seed will provide thousands of plants from a nursery. Direct seeding can be used. Plantings for oil are usually made in closely spaced rows of 90×75cm to 200×200cm.

Only leaf and terminal branchlets are used for distillation. Yields of herbage vary widely with species and conditions. Yields of 4 to 10t/ha can be expected in low-input systems. Yields in the first and second years will be reduced. Live pressure steam should be used for distillation.

Mint (Mentha arvensis)
Mint is a short-term perennial crop, and is typically cropped for three years, before replanting. The cut herbage yields an essential oil. This is then fractionalized (by simple chilling and stirring) into methanol crystals and dementholized mint oil. The market is very large. The crop benefits from a cool season, during which the herbage dies back.

The major points of interest for the crop are:

- After establishment the crop has total ground cover, and the soil is well bound by the dense network of stolons that develops
- During the cool season when the foliage dies, the ground is still protected by a thick mat of dead stems, and
- As a short-term perennial, the crop is well suited to intercropping in new plantings of tree crops.

The crop is propagated vegetatively by stolons. Sourcing of elite material is important. There are elite Brazilian and Indian clones available.

The crop is usually planted in rows. Row spacings vary from 40 to 100cm. Stolons are planted end-to-end along the rows. One hectare of established mint should provide planting material for 10ha. The crop can withstand extremes of termperature, though yields will be dependent on water availability during the growing season, and the length of the growing season. The crop is tolerant to saline soils. During the growing season, cuts are taken every eight to nine weeks, giving three to four cuts under typical conditions. The crop is cut when flowering starts.

Spices

A very wide range of spices can be grown in Africa, and all have high value. There are many tree crops (clove, pimento, nutmeg) with obvious long-term benefits to soil and water management. But these all take a long time to come into effect, and it is hard for new producers to develop commercial operations.

Two spices, both vines, are worth mention for the medium high rainfall areas as they are perennial, relatively early to yield (around two years) and can be grown on trees that provide shade and support.

Vanilla (*Vanilla planiflora*)

Vanilla is a crop of the humid tropics, and is typically associated with islands and maritime environments. Developing the expertise to produce good quality cured vanilla using traditional methods has always been the major constraint to the development of new producers. However, rapid processing techniques, following simple recipes, have been developed. These produce excellent quality with no requirement for traditional expertise. The crop is ideally suited to outgrower producers selling the fresh green beans to processors in the area.

The high price is set by a cartel (Madagascar, Reunion, Comoros) and is way above the costs of production. Therefore, the situation is perfect for non-cartel producers to supply the market with lower-priced, but still highly profitable, material.

Vanilla requires a tree crop for shade and support, a lot of surface mulching with organic matter. It cannot stand any soil disturbance by cultivation. It can, therefore, be cropped on steep hillsides with no erosion concern.

The crop is propagated vegetatively from cuttings, and fits excellently into established banana and coffee gardens where the coffee and bananas provide immediate shade. Ideal rainfall conditions are in the range 1750 to 2500mm/yr, but sites down to 1200mm/yr can be used if humidities and other conditions are good. Stimulation of flowering requires a short (three month) dry season, and flowers have to be pollinated by hand. Beans are ready for harvesting nine months later.

Pepper (*Piper nigrum*)

This crop also needs a support, and can tolerate and, depending on climate, may require moderate shade. It can therefore be grown like vanilla, on suitable shade/support trees. Mulch is also beneficial. Processing is very simple – blanching of the harvested berries, followed by simple drying. Pepper can give a range of products, in addition to the simple dried black and white pepper, such as green and red berries in brine, and a range of relishes. These products are all suitable for local cottage industries.

Pepper requires medium to high rainfall levels, ideally over 1750mm/yr. Lower levels can be tolerated if humidities are high and shade mulch levels are increased. Optimum temperatures are in the range 25 to 35°C, but it can tolerate temperatures in the range 10 to 40°C. Plantings are known up to 1500m. The complex of rainfall, temperatures, altitude, and humidity are all linked, and actual suitability of a site will depend on field assessment in marginal areas.

Oleoresins and other extracts

The essential oils are the steam volatile oils that are removed when steam is passed through the cut crop. If, however, an organic solvent (for example, hexane or ethanol) is passed through the crop, the steam volatile oils plus other chemicals are removed. Therefore, the extract has a closer resemblance to the original whole product in flavour and aroma. For many species, oleoresins have a very wide usage, particularly in the food industry where they are preferred for their uniformity and strength over the simple unprocessed spice. An example is the flower tuberose.

Tuberose (*Polyanthes tuberosum*)

Tuberose is a perennial bulb crop. It has a small, but very rapidly growing, cut flower market. Also, it is an exceptionally high priced perfumery extract.

The potential as a family crop cash crop would come from using the old, traditional wax block extraction process, in which open florets, picked daily, are placed between wax blocks. The extract migrates from the flower into the wax over 48 hours. The blocks are then sold. Such a process is completely compatible with small grower operations in remote locations.

Picking is labour intensive as the open florets need to be picked daily, but small areas can be managed with family labour and would give a significant cash income.

Dried flowers

The market for dried flowers is expanding rapidly. Two sectors of this market offer potential market niches. The particular interest of these products is that they can come largely from wild plants, making commercial use of the natural forest and scrub areas that need to be preserved, but which have to be given direct commercial value if the local inhabitants are to be able to retain them unaltered.

Pot-pourri

Pot-pourri comprises dried flowers, petals of flowers and other flower parts, fruit stones, and even desiccated fruits. The final mixed product is always scented with something, and the individual items may have natural colours or be artificially dyed. Customers place this mix in bowls in the house, where it functions as a pleasing alternative to aerosol sprays. The requirement is to have a content with interesting shapes, textures, and colours.

This is an ideal product in which to use material harvested from the wild. Being dry, storage and transport are not problems. The material must be robust, to withstand frequent mixing and handling.

Pot-pourri is a low-value, high-volume, product. The traditional market is the English-speaking countries, but it is starting to spread to continental European countries.

Artificial arrangements
Another section of the dried flower market purchases individual plant parts to make up into arrangements of artificial dried flowers. Again, the potential for new and interesting products is very strong, and the market is expanding rapidly. The products can all come from wild plants.

Dried vegetables

The market for fresh vegetables, airfreighted to Europe, is well-known, and well serviced by an established producer industry. The opportunities for small producers to tap into this market are very limited. Dried vegetables offer an alternative high-priced market that is well suited to small-scale production in dispersed areas. The demand from the catering and dried-soup trades is strong. Production of these vegetables would give farmers a high value cash crop that would be grown on a small area of the farm. Overall cash income would increase, thereby reducing pressure on other land resources.

Drying can be done simply. A small pilot drier to service a trial development could be purchased for US$5000. Driers are multi-purpose and can be used for other crops and products in different seasons.

The approach to development

The market-driven approach to crop selection has already been stressed. Within the context of market demand, crop selections are made on the basis of the potentials of the target location and producers. This should produce a selection of several possibilities.

Following on from this point, trials with both crops and processing facilities are essential. These must be linked to test marketing. The objective must be to move through the pilot commercial operations on a broad front covering all aspects. This ensures that when a producer is ready to progress to the next stage of a fully commercial operation, he or she is already in the market. The functions of pilot operations are first to identify the commercial crops, and then establish the technical requirements for production and processing, the financial cost and return framework, the marketing and investment structure. The final task is to take the product, and producers, into the market.

Pilot operations must be off-station, on producers' land. Distillation facilities can be under institutional control. The essential requirement is that producers are safeguarded against financial risk. A common method used is to guarantee to buy their crop at a fixed price. Thus producers start as contract growers. Careful appraisal of the overall operation, and modelling in advance, will provide price and cost information. In turn these will be used to define minimum targets for a viable commercial operation, and can be used as initial crop buying prices.

The markets for these high value crops are small. It is essential to work with specific companies from the earliest stage possible, and not rely on general market surveys for indications of price and demand. In the final analysis, the need is for outlets not information.

Community Development Case Studies

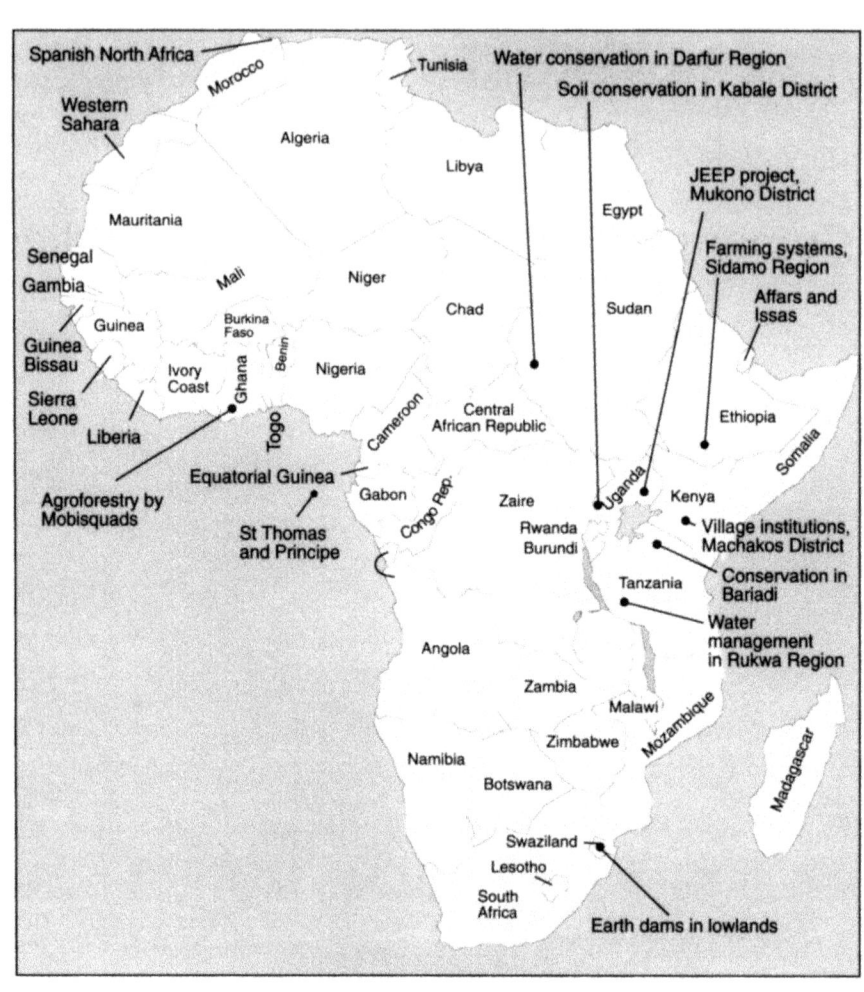

Community Participation

P. VEIT

Introduction

Several approaches to sustainable development based on work with communities are presented in this section. Throughout this section the emphasis is on telling the story of how a community organized themselves, rather than on details of each project. Each report demonstrates a link between participatory development and indicators of environmental and economic success. The responsibility for environmental management lies with the millions of resource users in rural Africa who work the land and make decisions concerning farming practices, resource allocation, and conservation. Throughout Africa, governments and development assistance agencies are recognizing the importance of community participation in the development process. Involvement in development planning often leads to ownership of project design. It also results in active participation in the implementation and management by local people, and helps ensure that local priorities are addressed.

There are different definitions of popular participation. To some, participation means local involvement in the implementation of externally designed projects. To others it means soliciting the views of local government authorities, introducing approved projects to development agencies, or visiting a few communities to discuss issues of local concern. Few activities have involved local people in decision-making, project planning, and formulating public policy.

Elements of effective resource management

Achieving community participation by developing an effective partnership in decision-making is a complex and difficult process. The examples presented in this section highlight important socio-economic elements of natural resource management.

Village institutions
Local management of commonly-held resources depends on united decision-making. In small-scale enterprises, such as collective farming or irrigation schemes, collective management is the responsibility of a users' institution which deals with system development, management, and administration. Successful institutions achieve active member participation through strong leadership, democratic decision-making, and sound management. These factors lead to institutional legitimacy, in the eyes of the villagers and the government. Thus, their effectiveness is increased.

Committed local leadership
Effective leadership of community groups is vital for successful development. The support and direct involvement of local leaders is crucial in

helping villagers to design and implement a prioritized action plan, assisting access to critical resources, introducing new practices and methods, establishing contacts with external development assistance agencies, legitimizing the development issues of local concern, and praising local efforts. Local leaders are also instrumental in establishing informal regulations about resource use.

The need for sound resource management practices

An increasing population density has forced farmers throughout Africa to cultivate marginal land or intensify production. Often the farming systems are unsustainable in the longer term. Inappropriate cropping methods lead to soil loss and yield decline. Generally, the production practices used are indigenous, or have been known for many years. Most have been changed, to meet on-farm crop needs in the most efficient manner with least risk, not for conservation of the resource base. In order to stop land degradation there is a need to identify low risk and efficient practices that integrate catchment conservation with production. A revised form of husbandry must appeal to the community institutions. It must build upon the social, technical, and ecological opportunities, and not increase demands for labour and capital.

Security of resources

Real or perceived security of resources encourages farmers to initiate conservation measures and undertake long-term investments. Therefore, the introduction of reform to increase land user control is likely to improve land husbandry. Even minor changes such as a shift from short- to long-term leases can achieve this.

Implications

Self-help is limited to farmer resources, organization, knowledge, and technical skills, and therefore external assistance is needed. Villagers know what their principal problems are, but not necessarily the best solutions. Linkages with external agencies are therefore needed to help farmers solve their problems, meet their priorities, and take advantage of their opportunities.

The community or village-cluster is a development unit that can be effective in resource management. Local labour and resources, combined with modest external assistance, and management by grassroots institutions with committed leadership, can foster sustainable development.

Devolution of administrative and political authority can promote community development by bringing decision-making close to the grassroots. In Uganda and Kenya this is being done by giving more power to districts. Most farmers feel the district is still too distant, because district development plans do not address village priorities. In Tanzania, political authority reaches village-level, but is hampered by the need for central government approval of community plans, policies, and by-laws.

Local-level plans and policies have more direct impact on the management of natural rather than national-level action plans, policy, and legislation. In Uganda, the responsibility for soil management lies with the district authorities, but most districts have endorsed a standard set of soil conservation by-laws. Thus, extension officers do not have the opportunity to develop appropriate site-specific land husbandry practices with farmers. The challenge is to effect changes in institutional functions and structures that will enable participatory development at the community level.

To be effective, government and fieldworkers need a new range of technical abilities and interpersonal skills to enable them to communicate and work with rural people. Training courses are needed to fit the present, and the next, generation for new roles. Currently, decentralized planning and implementation, community decision-making, conflict resolution, arbitration, and negotiation and other necessary skills are not taught to development specialists.

Communities often have the interest to lead the process of identifying local problems and opportunities, and implementing solutions. Unfortunately, few communities have the institutional and leadership skills to do this. Consistent support and training opportunities are needed from the government and development assistance community. This help is vital if village leaders and institutions are to achieve their important roles and responsibilities in grassroots development.

Conclusion

Policy changes and action are required to encourage and support community initiatives and local-level projects. The potential is enormous for increasing economic activity and promoting sustainable management through more local responsibility. The challenge is to help communities to identify their needs, and implement their plans for sustainable land husbandry practices, which bring an acceptable level of benefits.

Erosion Control in Machakos, Kenya

BARBARA THOMAS-SLAYTER, CHARITY KABUTHA, and R. FORD

Introduction

Twice in the history of Katheka Sub-location, Machakos District, its residents have been involved in the construction of bench terraces to control erosion and stabilize agricultural productivity. The first effort was coerced by the British colonial government, resisted by residents, and not sustained. The second effort began in the early 1970s and continues today. Fifteen traditional volunteer groups have been revitalized and have constructed over 20km of bench terraces and almost 100 check dams.

First, the effort was initiated locally, out of the realization of little hope for external assistance in the near future. Second, bench terracing is a known technology which is recognized as effective. Lastly, the effort is comfortably based within revitalized traditional local institutions, and it functions within acceptable social contexts.

Katheka's self-help erosion control

Katheka, approximately 11km^2, contains three villages with more than 1000 residents each. The area is classified as marginally productive semi-arid land with elevations from 1240 to 1500m. Porous soils are spotted with sparse vegetation, offering little protection against water runoff and resulting in massive soil erosion.

Following independence in 1963, the hostility towards coerced colonial-era conservation continued. Soil conservation practices stopped and erosion noticeably increased during the next 10 years. By the 1970s Katheka faced a crisis. The population was growing, but agricultural productivity was on the decline. The community became aware that resource degradation was out of control. The newly appointed assistant chief for Katheka encouraged the revitalization of *mwethya* groups, traditional voluntary self-help groups, primarily, but not exclusively, women. Each group has 25 to 35 members who equally share in co-operative labour. Members take turns working for one another, doing agricultural work, building and repairing bench terraces and digging cut-off drains.

The groups taught themselves bench terracing techniques by experimentation. Later, extension workers taught improved techniques. About 15 groups now build an average of 20km annually. In addition, they have begun constructing check dams in gullies, building subsurface dams, installing hand pumps, and developing markets for their hand-made baskets. They are also seeking funds for larger development projects and other new initiatives. It is evident that the *mwethya* tradition has been so successfully reestablished that the groups are the most visible and active of the local institutions. Indeed, they form the backbone of the village's resource management activities.

Although Katheka has effectively surmounted some of the technological and institutional obstacles to ecological sustainability on a local level, external challenges still remain.

Villagers are struggling with Nairobi business people to gain control of the area's abundant sand deposits in the Kalala river bed. Mining the sand is legal with an easily obtainable permit, but the long-term environmental impact is severe. These sand deposits retain water and provide a much-needed dry-season water source for the community.

Elements of effective resource management

The *mwethya* groups are succeeding for a number of reasons. Katheka's activities in resource management were born out of need when there was little hope of external assistance. In the early 1970s the effects of a decade of neglect were glaring. The villagers recognized that bench terracing reduced soil losses and construction of small dams increased water retention. They had little contact with the regional agricultural officers, water engineer or forestry officers over the preceding decade. They realized that if they were to prosper, they had to learn to define their own problems, set priorities for action, and find ways to mobilize local and external resources. Largely through the activities of the *mwethya* groups, Katheka's residents are united and have the confidence to take initiative in managing the local natural resource base.

A second reason for the success in Katheka is that a known and manageable technology was used. Bench terracing was a familiar technique, and its effectiveness in reducing soil erosion was recognized. Although Katheka's residents had heartily resisted the colonial authorities' coerced conservation programme, they were aware of its value. In the midst of the fervour of impending self-government in the early 1960s, coercive measures stopped and most terracing was halted. In 1973, when the villagers organized around the problem of declining agricultural productivity, it had been more than 10 years since anyone in Katheka had constructed a bench terrace. But the knowledge of terracing had not been forgotten.

Katheka's initial terracing activities were limited by a lack of local technical ability. But as this knowledge was shared internally, progress steadily increased. Each new success built on the visible achievements of previous efforts. The incremental approach has also enabled the villagers to learn and develop proficiency in the techniques of management, organization, and leadership to deal with the increased activity.

A third hypothesis for the sustainability of Katheka's resource management techniques is that they operate through viable local institutions and within an acceptable social context. Terracing activities do not disrupt social continuity, they fit well in the seasonal calendar of activities, and they lead to a relatively even distribution of benefits.

Realizing that single households alone do not have the expertise or resources to stop natural resource degradation in the sublocation, the residents of Katheka pooled their knowledge and energies to meet common environmental challenges. Rebuilding traditional *mwethya* self-help groups and mobilizing them in natural resource management practices is Katheka's greatest institutional accomplishment.

Several institutional elements have encouraged the flourishing of the *mwethya* groups. First, the broad political environment is supportive. The Kenyan government has expressed a renewed interest in rural development. The focus of its programmes has shifted to the district and has been backed by broad donor support. Additional donor support was attracted as a result of growing interest in the role of women in development. The Kenyan government's nationwide soil and water conservation programme, launched in 1974 and emphasizing self-help initiatives, provided national visibility for and legitimized the work of Katheka's *mwethya* groups.

A second element is the recognition of gender roles. In Africa, women traditionally assume responsibility for most farm work, and thus also for resource management. Participation in the *mwethya* groups is overwhelmingly female – 14 of the 15 groups elected women as leaders and women control the group's bank account. In organizing themselves for soil conservation work in this way, the women are acting in accordance with the traditional gender divisions of labour. However, men have recently begun to make significant contributions. Two of the 15 *mwethya* groups are mixed gender, and one is all male.

Third, the *mwethya* groups have flourished because of committed leadership. These leaders, generally elected by group members, have the respect and support of their constituency and enjoy considerable prestige and status in their communities. The combination of sound leadership and committed members brings access to labour, which is the single greatest asset available in most rural communities.

Fourth, the *mwethya* groups have benefited from the support and encouragement of traditional, governmental, and non-governmental village leaders, in particular, the assistant chief. The new assistant chief looked internally for ways to resolve local problems. He directed area residents toward self-help activities and arranged meetings to revitalize *mwethya* groups. His continued involvement has increased the status of *mwethya* groups within the community.

Implications

Kenya's district plans are an important milestone, not only for Kenya but for Africa. They suggest that decentralization can extend to Africa's rural areas, and to institutions that directly represent the people. Trends toward decentralization unquestionably spurred initiatives in Katheka. The Katheka example suggests that local labour and resources, combined with modest external assistance, and managed by viable village or grassroots institutions with committed leadership, can foster sustainable development.

Another implication which emerges from this study is that locally-based development technologies are more likely to be sustained if they are known to villagers or can be quickly taught, understood, and adopted by rural residents.

Katheka's development is limited, in part, because only the immediate local resources and institutional support are used. Katheka has been largely unsuccessful in establishing links with external agencies to help the community address their concerns. The Kenyan government has made efforts to decentralize planning, yet the Katheka example indicates that district-level planning can still be distant from village priorities. Mechanisms are needed to involve local communities more effectively in the planning process, and to integrate community plans into the regional context.

Conclusion

Katheka's successes suggest that the collective decision-making and actions of the community are keys to attaining sustainable development. In

Katheka, the community-driven approach has generated a social energy that has gained much more for the community than individual or coercive efforts could.

This is an abridged version of a more detailed paper listed in further reading under the *From the Ground–up* series.

Conservation in Bariadi, Tanzania

THE REV. H. GAPPA

'Better to light one candle than to curse the darkness'

This is not a scientific case study. Rather, it is a telling of the story of conversion on my part, and the continuing conversion of people in Bariadi District. My account is of an awakening, a change in mentality, a move to action.

I was assigned to Shinyanga diocese, Tanzania, in 1968. Shinyanga lies to the east and south-east of Lake Victoria. There is so much for a newcomer to absorb, with language, culture, and traditions. But it was easy to see the degradation of the land. People spoke of desertification. There was fatalistic complaining, but not much else. Tree seedlings were available. Some people were trying to reverse the trend. But the cutting of trees and overgrazing relentlessly moved on.

At one time, a delegation came to see me. Did I know about Padre George's attempt to set up a vegetable growers' co-op? Yes, I did. I also knew it had died. 'Could you help us start it up again? At that time we did not understand what he was trying to do. Now after 10 years, we do understand, and want to try again.' That taught me the importance of time and values.

Learning and preparing

After four years State-side, I returned to Shinyanga in 1977 to begin a new parish in the district of Bariadi. I had become better in assessing needs and responses, and recognized that the land issue was critical. *Kutunza ardhi* (to take care of the land) became a major objective for our parish. Trees were planted before the church building was constructed.

For five years I lived with the Catechist and his family, observing, asking, listening, discussing alternatives. I learned how difficult life is for ordinary families. We began planting. The first two years we had crops of maize – 11 and then 20 bags from our field. The vegetable garden, was planted in an old road bed. With sannhemp (*Crotelaria juncea*), double digging and lots of organic matter, the garden flourished. Living hedges protected our plot, seedlings grew into visible trees. We went public right from the start.

Our plot is on the main road, and people began to notice the change. They talked about us, and came for a look. We were on our way.

I began to reflect on the theology motivating our actions. Care of the earth is a spiritual matter, and most of the people with whom we work are deeply religious, be they of traditional, Muslim, or Christian belief. This element should not be overlooked as we seek methods and motivations in our struggle to save the earth and let it be a good place for people to live.

Homework at night included reading about organic gardening, which helped me shift from chemical and big, to small and organic. Books and magazines introduced me to green manuring and nitrogen-fixing legumes. Without much knowledge, without the clarification of purpose that would come later, we ploughed ahead. Our plot was going to regenerate! Hopefully this would encourage our parishioners and others to see and imitate. Or were we too late? Was the situation hopeless? But there must be something I could do.

Mobilization

'Show me' is a characteristic remark of many rural folk. 'You do it first. If it works, I might try it. Show me!' And I had never had a classroom course in agronomy, agroforestry or anything of the sort. But we talked about things in meetings, at homes, in seminars, in church. At that time planting trees was not a priority in our neighbourhood. Some would even chide me, '*wazo mzuri, Padre, lakini haiwezekani*', (a good idea, but no go).

For farmers, a favourite religious ceremony is the blessing of seeds at planting time. No longer was this a quick ceremony. Preparations for it now took four Sundays. It became a skit showing the pressure on the land, talks on good farming practices, the benefits of trees, problems of overgrazing, hope versus fear and fatalism, the need for change. One Sunday homily was a tour of our land to show what we are doing, what works, and our plans. Finally came the blessings.

Our objective was threefold – plant trees, feed the soil, and make permanent beds for the vegetable garden. Sannhemp and leucaena trees thriving right outside the church showed what we were talking about. A new theology evolved. When Masanja dies, he goes for his final evaluation. When St. Peter, the Keeper of the Gates, finds out that Masanja is from Shinyanga, he asks only two questions: how many trees did you plant? And how many did you cut down? Poor Masanja is in a state of shock. Simple arithmetic determines whether Masanja goes through the pearly gates or downstairs forever! Again, when Milembe is preparing to join the church, I ask her: 'why do you want to be baptized?' The usual answer is 'to save my soul'. Imagine Milembe's consternation when it is revealed to her that she will be baptized, not to save her soul, but to plant trees. All of this is tongue in cheek of course, but they laugh and the point comes across.

We asked each homestead to plant five trees a year. The planting is the easy part. Caring for them is the test of resolve. One day, an elder said to me, 'Your teaching is different – you not only say, you do'. We were given more land, and we could demonstrate boundary planting, rows of mixed species for agroforestry, revive waste lands. Advertisement increased with small plots in several village outstations.

We had all the familiar problems. Lack of time, energy and funds, termites, drought, five to six months dry season, poor soil, people gathering fuel from the fields (cornstalks or thorns placed to protect seedlings), and the sociological confusion resulting from something new on the scene. There was the never ending battle with livestock – explaining to herder, chasing goats, settling fines, and the inevitable court cases. *Sio rahisi* (it isn't easy)!

A major contribution came when Hashi arrived on the scene. *Haifadhi ardhi Shinyanga* is a government programme called the Greening of Shinyanga. The young district manager has proven to be a dedicated professional who has the skills to make things happen, working with officials and villagers, producing seedlings and woodlots. It is a real pleasure to work with him and watch the results. People are now ready for such an endeavour.

We noticed the change about two years ago. I cannot say that every home now has 60 trees, like the plan called for, but few have none. Strangers come by to look, and say *hongera* (congratulations), and tell us about their endeavours. Some are planting trees by the house, others small woodlots, others just giving local species a chance to grow. Seldom do livestock invade our fields now. In fact, in some places we welcome herders to mow the grass. Some people figure that in five years our district will have turned the corner in reforestation.

Bariadi is certainly not a jungle, but it is a far cry from the inevitable wasteland that we could do nothing with 10 years ago. It will take more people, time, field demonstrations, and conversion before trees will be utilized as an integral part of that agro-silvi-pastoral balance that we all dream of. And we feel good when we are told that our little parish plot had something to do with it all. It is better to nurture growth than to watch death.

Comments

There are, I am sure, other and better ways to accomplish a work such as this. A critic could find many faults. I had to squeeze ecology into busy days filled with pastoral care. Nonetheless, I think it safe to say there is more going on than when we started. It was not the great leap forward, not a sudden grassroots movement. I think people were not ready for something like that then. We tried to show, gently and slowly, that degraded land could be regenerated. The people watched our every step. We realized it is only a drop in the bucket, a first step on a long safari. But that first step has been made.

Factors that worked for us:

- For motivation, two things stand out. Dead land coming back to life with good crop production, and cash sale of firewood. Money talks!

- Patience, taking time needed, over 10 years
- Language and credibility – how one relates to people
- Stubbornness of the innovator
- Visibility and ordinariness of the project
- Going public. Talking about the project frequently
- Showing interest, giving support, and believing in people
- Willingness to be unconventional and take the consequences, and
- Willingness to start with what we had, but to start.

Obstacles we have encountered:

- A small percentage of people responded positively. Acceptance of a new value system takes time
- Insecurity of land tenure
- Livestock, their owners and herders. This is the most discouraging factor. There is still no local model to demonstrate the long-term benefits of trees to livestock, and that it pays to wait for trees to mature, and
- Lack of resolve to nurture trees to maturity.

Community Soil Conservation in Kabale, Uganda

E.M. TUKAHIRWA and P. VEIT

Introduction

In Nyarurembo Sub-parish, south-western Uganda, several farming communities have been continuously cultivating steep hillslopes for more than 50 years, without significant losses in productivity. With the assistance of an enlightened agricultural extension officer, the farmers develop and implement site-specific soil conservation plans, including the use of band, strip, and bench terraces. Four elements contribute to the successful experience in Nyarurembo Sub-parish. First, since 1939 district soil conservation by-laws have legitimized practices and empowered local authorities. Second, the people recognize the need for soil conservation. Third, the good relationship between the local extension officer and Nyarurembo farmers has facilitated the design of soil conservation strategies. Fourth, farmers operate under a customary land tenure system which fosters a sense of security.

Terracing for improved agricultural production

In the mountainous south-west corner of Uganda, Nyarurembo is dominated by hills 30 to 160m above the general altitude of the area with peaks around 2000m above sea-level. Soils are generally fertile, but shallow and highly erodible. Originally, the hills were covered with open canopy montane rainforest and the lowlands with bamboo and small shrubs and grasses. Today, most land in Nyarurembo is cultivated or in fallow. No natural forests remain. As the population increased the need for arable land forced the farmers to cultivate the steeper hillsides. Over-cultivation and inappropriate farming methods were denuding the hillsides, causing soil erosion and contributing to declining agricultural productivity.

The farmers in Nyarurembo Sub-parish have practised soil and water management for 50 years. The most obvious practices are the terraces, band, strip, and bench on the hill slopes. The types of terracing practised depends on ecological and socio-economic factors.

Band terraces are contoured bands of land, about eight metres wide, that encircle the hill and are alternately cropped and fallowed. A one metre wide path is cut between every two bands. The natural vegetation growth on the fallowed band captures soil that erodes from the cultivated band above it.

Strip terraces are also contoured bands of farmland eight metres wide that encircle the hill. Between each band, the farmers plant a one metre wide strip of napier grass (*Pennisetum purpureum*). The grass strips trap soil that erodes down the slope. Over time, a natural terrace builds up in and above the strip. Farmers continuously cultivate the land with two crops a year.

The terraces are knocked down every eight to nine years when the accumulated soil threatens to collapse the terrace, and the area immediately below the grass strip becomes bare of soil and agriculturally unproductive. The soil is redistributed and the grass strips are redemarcated four metres above and below their previous location.

Bench terracing is the most labour-intensive form of terracing practised, and the least common. Bench terraces consist of permanent, level steps cut into the hillside along the contour. The width and depth of the steps vary with the available labour, hillslope, and soil depth. In Nyarurembo they are usually less than 1.5m wide and deep. The land is continuously double-cropped.

Elements of effective resource management

While Nyarurembo's farmers have effectively addressed the threat of soil erosion, the problems of land shortage and land fragmentation apply pressure to the resource base. These make it more difficult to adopt improved agricultural techniques. Some important factors emerge from the experience in Nyarurembo.

First, the recognized need for soil conservation on hillside farms by the farmers. This, coupled with the proven effectiveness of terracing to control erosion and maintain soil fertility, has helped to ensure that terracing practices have continued. The high population density has forced the farmers to cultivate steep hillsides, but they understand the relationships between soil erosion, soil fertility and agricultural production.

Band, strip, and bench terracing are locally understood to be effective soil conservation techniques, and are generally credited with the maintenance of the area's agricultural productivity. During the period of more than 50 years that terracing has been undertaken the practices have been absorbed into the agricultural strategy of every hill farmer, and into the prevailing socio-economic value system. This suggests that the soil conservation practices are sustainable, and will continue.

Second, the Kabale district by-laws have been a driving force behind the effective terracing practices. The legislation legitimizes the importance of soil conservation and empowers the local authorities to implement and enforce practices approved by them, not necessarily the specific practices mandated by the by-laws.

The district soil conservation by-laws regarding terracing have never been implemented or enforced, because the agricultural officer does not recommend it, whereas the alternative terracing forms are approved of, and endorsed by the local authority.

The legitimacy of the local authority in farmers' eyes has also enabled them to impose sanctions not prescribed in the by-laws, such as the five-year closure of Nyarurembo Hill. The closure was particularly severe, but community vigilance of terracing practices has increased.

Another force behind Nyarurembo's success is the efforts of the agricultural extension staff. The agricultural officer's effectiveness can be attributed to his genuine interest in the welfare of the people from his home area, and to the good relationships he has developed with farmers. He also has sufficient confidence in his knowledge of agriculture and local socio-economic and ecological matters to occasionally work outside the by-law mandate.

The agricultural officer is also an important link between the farmers and local authorities and development agencies. He is effective in this role because of the good working relationships he has developed with the local chiefs, councils, and a range of other civil servants. Through his contacts, the agricultural officer has effectively mobilized external resources to provide critical inputs for soil conservation.

The extension officer recognizes farmers' participation in decision-making is critical to effective soil conservation. He helps to organize meetings of hill farmers to develop hill-specific terracing strategies that best meet local needs and constraints. He works to reach a consensus on critical terracing actions, and to co-ordinate the terracing and related activities.

Another element contributing to the success is the farmers' perceived security in the land. The current customary land tenure system provides people with access to, and control of, land. It also determines the right to inherit, rent, or sell the land and its resources. This security in their land and customary tenure has encouraged sound land use and management with long-term goals. It has discouraged unsustainable farming practices, and the mining of natural resources for short-term objectives.

Implications

The success of the Nyarurembo programme suggests that this experience can be fed back into national planning, so that the approach can be applied on a national scale. An overview of the Ugandan government's environmental policies shows that studies are being made to address natural resource management and local development issues.

The five-tier structure of resistance councils and committees which extends down to community level is one of the most visible of three initiatives, and further devolution of political and administrative authority is desirable and probable.

There is clearly room to reform the present legal land tenure system, based on the unloved Land Reform Decree of 1975, in favour of the traditional land tenure system. The sense of security from the customary land tenure system in Nyarurembo has encouraged sound land use.

The ability of local authorities to take a relaxed approach to enforcing legislation is clearly important. The district recommendations did not fit. Thus locally acceptable variations were substituted, and adopted, because they were familiar and known to be effective.

This study shows the crucial role of effective advisory services through agricultural officers and extension officers who are local people, known, trusted, and respected. National training programmes should emphasize communication skills and train local people to work with their peers.

This is an abridged version of a more detailed paper listed in the further reading under the *From the Ground–up* series.

Agroforestry by Mobisquads in Ghana

C. DORM-ADZOBU, O. AMPUDU-AGYEI and P. VEIT

Introduction

In Goviefe-Agodome, Volta Region, a local self-development co-operative initiated by the government has successfully turned land considered infertile into productive farmland through various agroforestry practices. The co-operative has emerged as the most active community development institution.

Four elements contributed to the success of Goviefe-Agodome's community development efforts. First, the actions are designed, implemented, and managed by a local organization acceptable to the community and with the support of local leaders and institutions. Second, the agroforestry efforts yield immediate financial and other benefits to the co-operative members and their households. Third, the resource management activities are locally sustainable, involving familiar techniques and practices. And last, the community benefits from its accessibility to major urban areas and has received much development assistance.

Agroforestry by mobisquad

The government of Ghana, which came to power in December 1981, realizes the importance of popular participation in community development. It also accepts its limited capacity to reach the large rural population. To provide a vehicle for mass mobilization, the government called for the establishment of several village-based so-called revolutionary organs, including the 31st December Women's Movement and the village mobilization squad (hereafter mobisquads). Operating throughout Ghana, many are involved in activities that have improved local socio-economic circumstances while maintaining the natural resource base. This study examines the development initiatives of the Goviefe-Agodome mobisquad, organized in 1986.

Goviefe-Agodome is situated in Hohoe District, Volta Region, 160km from Accra, and is one of five traditional *Govie* settlements, each headed by a village chief selected by a council of elders. The council, which helps the chief decide local matters, is made up of the leaders of the village clans. The queenmother is the traditional leader of women villagers. The *asafo* company organizes all young village men into communal work parties.

Population growth of the Goviefe communities is pressurizing available land in the area. Farm holdings have been fragmented, fallow periods have been shortened and cropping periods extended. Soil erosion and water runoff have increased, and crop yields have declined.

When more than one million Ghanaians were expelled in 1983 from Nigeria and Togo, the influx added more pressure on a weak economy and fragile resource base. In 1984, the government's National Mobilization Programme encouraged the formation of village mobilization squads for purposes of community development and national economic recovery. Groups received government support for some projects. But most raised funds with income-generating activities, such as cash cropping, gari processing, and fuelwood and charcoal production.

In Goviefe-Agodome, six of the returning Ghanaians organized a local mobisquad. With the political backing of the government, they worked to gain the approval and support of the village leadership and to recruit local leaders, including the village chief, queenmother, and *asafo* company. In 1986, the 41 member mobisquad's established a 4.8ha communal farm of cassava, maize, and cocoyam intercropped with teak and leucaena trees, on land previously considered wasteland. The mobisquad divided most of the profits equally among its members, and launched two community service projects – an improved latrine and a clinic.

During the next three years, the mobisquad grew in membership, and in the range of activities undertaken. In its first four years of operation, it developed a 37.6ha agroforestry farm, planted 19.2ha of cotton, and transplanted about 17 000 tree seedlings. From these and related activities, the mobisquad netted more than 3.28 million cedis (US$9850.00 today).

Local needs, problems, and untapped opportunities remain despite the many mobisquad successes. Although the mobisquad has established an executive committee to co-ordinate its activities, the squad may need further training on institutional growth and administration in the future. Of particular concern are two points. First, few Goviefe-Agodome farmers, mobisquad members and non-members, are executing agroforestry or other resource management practices on their own farms. Second, few neighbouring communities have developed viable mobisquads or involved their residents in sustainable community development efforts. Some training in intercropping and other resource management techniques would speed adoption of these practices on private farms in Goviefe-Agodome and nearby communities.

Elements of effective resource management

Four elements contributed to the success of Goviefe-Agodome community development efforts. First, the mobisquad established itself as a viable local institution. It enabled villagers to design, implement, and manage their own development efforts. The mobisquad has developed by gaining legitimacy as an institution and by taking on new challenges. Support and involvement of the local leaders were crucial to the establishment, legitimacy, and early achievements of the Goviefe-Agodome mobisquad. Village leaders worked with the mobisquad to donate communal land, facilitate weekly meetings for organizing mobisquad activities, encourage and praise squad efforts, and establish contacts with external development assistance agencies.

Traditional institutions were weak and essentially dormant when the mobisquad was set up. Local leaders and institutions view the mobisquad as an effective means of filling an institutional void left vacant by the traditional groups, thereby better meeting community needs. Thus, positive working relationships and collaborative efforts exist between the government-sponsored mobisquad and the traditional village institutions.

What has further helped the mobisquad's community development efforts is the direct link between planning and action within the group. Mobisquad members participate directly in planning and implementing most efforts, although non-member village leaders and other groups may also be involved.

The second root of Goviefe-Agodome's success was that individuals recognized the benefits of membership. While the contributions to the mobisquad and community have increased, individual member profits remain high. These are a significant incentive to join the mobisquad. As mobisquad services to the membership have increased, significant non-economic benefits have emerged. Members have access to low-interest loans from the mobisquad bank account, and to labour for agricultural, house-building, and other activities.

The third element was the fact that the principal income-generating activities of the Goviefe-Agodome mobisquad involve practices and techniques familiar to its members. Mobisquad efforts emphasize the production of cassava and maize, for which the knowledge and resources are known and locally-available. The concepts of self-help and work parties are familiar traditional arrangements.

Similarly, treeplanting and management are known, although less common. One important aspect of the local treeplanting activities is the mobisquad nursery, which is needed to produce such a large number of seedlings. The nursery is successful, in part because the techniques are sustainable locally, and are easily shared through farmer-to-farmer training.

The fourth element contributing to Goviefe-Agodome's success was the existence of links to external development assistance. The proximity and accessibility of Goviefe-Agodome to urban areas offer residents various socio-economic opportunities. Some residents have acquired skills, or have links to development assistance organizations, of potential benefit to the community. Publicity about the Goviefe-Agodome mobisquad has increased the number of visits by both government agencies and the development assistance community. As a result, numerous development assistance agencies have supported the community's efforts through technical advice, training, and resources, including funds.

Implications

The Goviefe-Agodome experience shows that the collective decision-making and power of the community regarding resource use are the keys to sustainable development. The government has worked to democratize political decision-making, decentalize power, ensure popular participation in the development process, and revive traditional communal spirit and self-help at the local level. In many cases, government policies are backed by legislation, an institutional infrastructure that reaches the community level, and programmes for local involvement in the planning and implementation of development.

These implications lead to recommendations for government and development assistance agencies. Although Ghana appears to be on the appropriate path to decentralization, the government's commitment to decentralization and popular participation requires appropriate grassroots institutions capable of motivating the people to take responsibility in community development and local resource management. In the short term, communities would benefit from technical assistance to catalyse this process. In the long term, village leaders could be trained in appropriate organizational skills and management techniques. Of equal importance, district environmental planning teams need to be trained to interact with villagers, to grasp real needs and opportunities, and to integrate local priority needs into district development plans. These teams could also act to help communities prepare sub-district development plans on which the district development plan could be based.

Another recommendation emerging from the Goviefe-Agodome experience is that development assistance agencies should identify and work with the most viable existing village groups. In Ghana the mobisquad is often the most viable local institution. In many communities there are other organized groups that could play an active role in development.

Conclusion

Throughout sub-Saharan Africa, governments and development assistance agencies are recognizing the importance of popular participation in the development process. The fundamental role and involvement of local leadership and village-based institutions in community development will become clear.

This study identifies some prerequisites and potentials of viable externally-initiated and sponsored village institutions. Government's political backing can significantly strengthen a local institution by legitimizing the group and its activities. Coupled with the support of local leaders and other village institutions, such groups can mobilize the labour and resources of members and non-members toward community development.

The roles and responsibilities of governments concerned with developing a favourable political, economic and social climate for rural communities are clear. Ghana's newly elected 110 district assemblies have the potential to reach the communities, and to involve them in district-level development and environmental decision-making. Other nations in sub-Saharan Africa with similar objectives are closely monitoring the outcomes of Ghana's decentralization process.

This is an abridged version of a more detailed paper listed in further reading under the *From the Ground-up* series.

Water Harvesting in Darfur, Sudan

YAGOUB A. MOHAMED

Introduction

The inhabitants of Wadaa village, in Darfur region of Sudan, paid great attention to the protection of natural resources through their local institutions. This case study illustrates the capacity of local organizations to adjust themselves to outside influences while maintaining their traditional role. The dominant strategy followed by Wadaa farmers is to avoid risk in the semi-arid environment and achieve food security. The move from rainfall farming to the form of water harvesting known locally as *trus* cultivation is an attempt in that direction.

Trus cultivation in Wadaa village

Wadaa village is the seat of a tribal leader, with a current population of approximately 8000. Each tribe in Darfur claims a certain territory as their own, which is then divided among the different sub-tribal sections and clans. Wadaa village lies on the borders of two topographically different areas, the sandy Qoz lands to the east, and the clay plains to the west. Rainfed agriculture on sandy soils and animal raising are the main activities, designed to satisfy first subsistence needs, and then provide for cash income.

The loss of *Acacia senegal* gardens, decline in soil fertility, and reduced rainfall in recent years has resulted in low yields and increasing uncertainty with regard to family food supplies. This stimulated the search for indigenous solutions to the problems.

Some innovative farmers began to tap the potential found in clay soils, which previously had only been utilized for small vegetable plots. Their success in using the technique of *trus* cultivation, a traditional water harvesting technique, combined with repeated years of drought and crop failure, provided a strong incentive for others to utilize clay soils. By 1983, nearly eighty per cent of the villagers possessed two farms; one in the sandy soils and the other under *trus* cultivation.

Trus cultivation is a simple technique that requires selection of a gently sloping site, a water spreading system, construction of embankments, and selection of a suitable crop. Plots are constructed, bounded on three sides by embankments while the top is left open to accept runoff. The embankments are made by hand using simple tools and are 30 to 40cm in height. Because of the labour input required for construction of the embankment and for farming operations, plots are generally small.

The shift to the use of clay soils also has positive impacts on the degraded sandy soils. It reduces further expansion of cultivated areas, and permits some farmers to abandon some fields to *Acacia senegal* growth.

Despite the success achieved by Wadaa farmers, there are technical and socio-cultural limitations that reduce the effectiveness of *trus* cultivation. First, the design of the system as practised today is based on trial and error. Substantial resources are wasted while alterations are made before reaching the best field alignment.

Second, greater involvement in *trus* cultivation has complicated the issue of landownership and created a trend toward privatization. Land is increasingly becoming a commodity, and this has already begun to restrict the participation of the poor. This problem of land tenure is critical and may impede the implementation of the strategy supported by the Darfur government to achieve a large-scale shift of farming activities to clay soils.

Third, *trus* cultivation has resulted in more production of vegetables for sale. Cash cropping in the clay soils does not improve local food security because of the continued high risk of crop failure.

Core elements

The following elements contribute to the success of the resource management activities in Wadaa. First, the local institutions showed a capacity to adapt to new roles. These arose from the failure of the modern institutions to respond to local needs and priorities. The modern and traditional institutions co-operated and their functions complemented each other. Through this interaction, a modified form of traditional organization was created, involving a division of responsibilities.

Second, the culture of participation prevailing among Wadaa inhabitants made it easier for grassroots groups to mobilize local inhabitants to participate in resource management activities. The sense of belonging to the tribe, and the evident degradation, created an awareness of dangers the community might face. This strengthened participation in collective action through familiar cultural models.

Third, a combination of capable organizations and influential and innovative leaders guide the process of resource management. The traditional authority still controls all decision-making power, but, in order to be effective, the so-called modern organizations work through tribal authority to the central government. This achieves co-ordination and integration of modern and traditional institutions.

Fourth, the concern for food security plays a critical role in resource management. Local planning centres on survival strategies, and is reflected in diversification and the shift to *trus* cultivation.

Fifth, the homogeneity of the population in the area fostered a spirit of co-operation and the smooth allocation of land. Because new settlers were of the same tribe, they enjoyed the same right to acquire land.

Sixth, the worsening of harsh conditions and the spread of desertification combined with the visible success of *trus* farming in the clay soils attracted more villagers to use the system.

Seventh, the technical knowledge of *trus* cultivation is common to many farmers and easily acquired by new practitioners. The design is simple and easy to construct. All that is required is some sense and experience to determine the gradient of the slope and hence the size of the embankment.

Implications

This case study has demonstrated the critical role of local institutions in the management of common resources in response to felt needs. It also highlighted their ability to mobilize the local people, and to guide them to achieve change in harmony with local culture and customs. However, these local institutions are often ignored by government which prefers new organizations to guide village development.

There is a need to understand and record customary rules related to land use and land tenure. Land tenure issues emerge as one of the factors which may hinder the expansion and development of new types of farming. Often the land selected is said to be community-owned. But as land value increases with the introduction of improvements, claimants appear who assert long-neglected claims to the land. There is a need for measures which will increase the access of poor households to *trus* land. Despite the 1970 government declaration that all unregistered land was government property, various customary practices have developed and are in use, such as those in Wadaa. Efforts need to be made to document clearer government policies.

Further, more government attention needs to be given to local initiatives. It should be recognized that these traditional techniques represent the accumulation of knowledge and experience, and can be starting points for adaptive research for local conditions. Planners and extension workers should appreciate that locally-designed strategies hinge on risk mitigation. With that understanding, they can work with local farmers to adapt the indigenous knowledge and their technical experience to solve problems.

Finally, more attention should also be given to the roles played by traditional and local institutions in the development process. This culture of participation is not yet tapped by government authorities. This case study revealed that local participation in self-help projects was crucial to successful natural resource management.

Joint Energy and Environment Projects (JEEP): wood energy conservation by participatory activities

RUTH KIWANUKA

Introduction

Joint Energy and Environmental Projects (JEEP) is an indigenous voluntary organization founded by a group of Ugandans who wanted to address environmental problems through joint action with people living in rural areas. The founder members identified deforestation and soil erosion as important processes impacting upon the environment. They decided to work with people to combat environmental degradation.

Deforestation has given rise to fuelwood shortages around villages, thereby adding to the burden of women. More time must be spent upon collection of firewood. Ninety per cent of domestic fuel is wood so there is a need to encourage the re-establishment of sources of supply and to improve the efficiency of energy use by improved practices for charcoal preparation and the burning of fuel at home.

JEEP's work

JEEP is developing a participatory development strategy to deal with the changes needed in respect of domestic energy supply and use. The organization works with fuelwood producers, as well as users. For example, a 1991 seminar with charcoal burners has led to a resolution to form an association for training and information sharing. The main work, with communities, varies according to culture, tradition, religion, and stage of development, but the following principles apply to all targeted groups:

- Everyone is encouraged to put forward ideas and opinion
- All available media are used
- Complex issues are explained by reference to local experience and knowledge
- Farmers are given comprehensive information and shown how to change their practices, and
- Advice and practical support is given to projects decided upon by community groups.

It is recognized that people get frustrated and may ignore proposals if they did not participate in the planning. Local participation is therefore required at all stages of project identification and implementation.

JEEP's messages

No meaningful development can take place if the people's cultural values, practices, and traditional beliefs are ignored. The history of cultural experience has brought about a close relationship between culture, development, and nature conservation. The linkages and lore are well-expressed in folk tales, songs and rituals which are passed on to the next generation. The use of theatre extends the traditional forms of communication, helps promote education and disseminate information in a lively manner, and people relax and learn in the company of each other. By running open theatre with opportunities for popular feedback indigenous knowledge can be explored, and local skills and talents improved. Theatre is used for development, and makes people appreciate the role of culture in national development. In selected villages in Mukono and Mpigi Districts local actors are employed to stage amateur drama using local songs, sayings, dance and poems. Experiences gained through co-operation between JEEP and farmers is woven into the material presented. For example, information on fuelwood shortage and ways of overcoming the problem have been dramatized. The effect is to stimulate farmers into positive action. After each performance an outline of the drama in comic strip form is distributed to every member of the audience. JEEP has resolved to take a further step with plans to work with NGOs, line ministries and theatre groups to organize drama festivals on culture and the environment.

The success

JEEP has assisted institutions with the construction of fuel-efficient stoves. The women in the villages have experienced improvements in their circumstances through the adoption of improved fireplaces and kitchen management. Village contact groups have also planted trees according to their priorities, for fuel, fodder, fruit, poles, timber, and as wind breaks, within their food gardens and on sites unsuitable for cropping. Advice has been given on how to produce seedlings and manage trees. Women participate in farmer-to-farmer exchange visits arranged by JEEP. The farmers' consensus is that they learn a great deal about agroforestry from this form of extension. Work in five pilot villages is now well established, and 75 000 tree seedlings have been planted within three years, with a survival rate of sixty-eight per cent.

Traditional Water Management and Irrigation Systems in Tanzania

OPHELIA MASCARENHAS

Introduction

This study reports a successful effort in building on local management of natural resources. It focuses on a traditional water management and irrigation system in Msanzi village, but it has implications that equally apply to the management of other natural resources. It demonstrates a process whereby communities become aware of ecological problems that threaten their livelihood and respond to them.

Irrigation practices in Msanzi village

Msanzi village is in Sumbawanga District of Rukwa Region in south-west Tanzania, 1300km from the capital, Dar es Salaam. The village is endowed with a natural valley at the confluence of two rivers. Lack of export crops and a weak economic and social infrastructure make for one of the least developed regions of the country.

Before 1933, the villagers grew rainfed crops in the upper valley and used the residual moisture to grow dry season crops. In 1933, one entrepreneur discovered that it was possible to use the waters of the upper valley to irrigate the lower valley and the dry plains below to cultivate dry season crops. The system consisted of a simple weir for intake, a main canal, and subsidiary furrows and ditches to divert the water from the canals to the fields. It was devised, maintained, and operated entirely by the people themselves through a very informal management system.

In the mid-1980s severe soil erosion set in and made the irrigation system for the lower valley and plain inoperable. The upper valley was not sufficient to meet the increased demand for food and cash. The villagers therefore decided to organize themselves to rehabilitate the system. Since a new technology was needed to prevent a repetition of the previous catastrophe, it was decided to use a partnership system in which an outside agency, Rural Development Programme (RUDEP), provided technical advice, while the village provided labour.

The Msanzi case study shows that traditional irrigation systems have limitations on the extent to which they can be extended in response to population and economic pressures. When these pressures lead to a collapse of the system a new technology is required and external assistance may be needed. This external involvement can create dependence and can fail to satisfy indigenous expectations. In the Msanzi case, the villagers had expected a much larger area to be available for irrigation. The irrigation committee, consisting of five members, is planning to expand the system using skills learned from the RUDEP expert. If this is accomplished, the irrigation system will be a true success. It will show that the peasants in Msanzi can protect their irrigation system using modern technology without external assistance, once the technology has been acquired.

Core elements

The Msanzi water management system is an effective use of natural resources that enables the villagers to overcome some of the deficiencies of rainfed agriculture. It is important to understand the principal factors contributing to the evolution of the system.

First, the villagers' predominant concern has been to minimize the risk associated with the unreliable rainfall. The villagers realize that the valley, and the plain beyond the confluence of the rivers, are prime resources with plenty of water and fertile alluvial soils. The people consider cultivation of the valley and plain as essential to increase food supplies and obtain incomes that they can no longer get from the primary crops.

Second, the villagers are using the available labour efficiently. Crop production during the rainy season cannot be expanded because of the labour requirements during a fairly short agricultural season. Dry season cultivation uses labour available at that time. Food supplies and labour use are spread more evenly over the year.

Third, increasing need and new opportunities are encouraging the adoption of irrigation practices. The traditional form of land use was based on shifting cultivation and a fallow of 10 to 12 years. Land is becoming more scarce and as a result the fallow is shortened and productivity of the land has dropped. At the same time there is a need for diversification from the main cash crop of maize.

Fourth, the development of irrigation in Msanzi has depended on leaders, who by their successes have disseminated the idea of others. At present, leadership is being provided by a group of better-off farmers who are experimenting with new techniques and new crops. This leadership is not resented in the village, in fact it is respected, and contributes a great deal in terms of advice, demonstration, and help. The villagers are more likely to listen to such leaders who are part of their community, particularly as there is no compulsion and it is seen that the leaders do not personally gain from the spread of a new technology.

Fifth, the Msanzi experience benefited from the existence of a framework for local participation in decision-making and management. In 1987, when there was a threat to the dry season cultivation, there was already a framework of institutions for organized action. The village assembly and village council provided a forum for discussion and decision-making. The

10-cell institution, a group of 10 households with one elected leader, implemented communal projects. Irrigation committees had to be created to supervise day-to-day management of the project. This framework was recognized by the villagers as the legitimate authority in the village, to plan and implement new projects. Actions were taken with the participation of the villagers and at the request of the villagers through their village assembly and village council.

Sixth, the Msanzi project shows the importance of having realistic goals and the flexibility to make adjustments. In this case the choice of crops grown and flexibility to intensify one or another according to the needs of the household. Flexibility was also apparent in the way that the irrigation system dovetailed into the rainfed system in terms of labour distribution and timing of activities.

Implications

The elements identified above have implications for policymakers, donors, and local government workers, for channeling development assistance for the promotion of natural resource management in Tanzania. This has to be seen in the context of the country's irrigation sector and its decentralization policy. Government policy envisages a three-pronged approach to irrigation development – through irrigated state farms, village irrigation projects, and traditional smallholder irrigation schemes. The Msanzi case study gives a different perspective on the potential of community-based and managed irrigation systems. There is a need therefore to allocate more resources to improving village-level irrigation schemes.

Although the predominant trend of centralized planning in Tanzania remains strong, efforts have been made to decentralize decision-making and implementation of local-level projects. One lasting obstacle to local-level management is the question of funding. Although currently districts are allowed to raise their own funds, in some regions where resources are largely undeveloped, the sources of income are limited.

There is also the realization that too many government-induced strategies have eroded the former spirit of self-reliance. They have created dependence on the government and aid agencies to resolve problems. Under such circumstances, the efforts of communities to work out sustainable development strategies through communal endeavours, perhaps with limited partnership with external agencies, offers an alternative that is appropriate and worth adopting.

Six recommendations emerge from this study. First, it is important that development strategies, to be sustainable, build on local existing systems. New technologies should be introduced when they build on the existing system. Since there are serious constraints on capital, development projects should concentrate on strategies which require little capital investment, but which have the potential to increase household production.

Second, it is important to incorporate indigenous knowledge into projects. The Msanzi case study showed that the irrigation was successful because it incorporated indigenous knowledge about natural resources and the potential, and limits, of rainfed and irrigated cropping. The RUDEP philosophy of self-help, local initiative, and education helped the people of Msanzi to incorporate external assistance into a traditional system.

Third, it is important to foster community participation and management. Traditional irrigation schemes are characterized by a high degree of the people's involvement in all three phases of the scheme – planning, implementation, and management. In Msanzi, locally-based decision-making and management is one of the critical aspects of success.

Fourth, there is a need to focus on improving traditional systems. The Msanzi case study shows that traditional systems have limitations. Traditional irrigation can be improved with external advice, scientific knowledge and small inputs.

Fifth, assistance can be channelled through local innovators. The Msanzi case study has shown that such entrepreneurs can be very useful in disseminating new technologies. Unlike the agricultural extension service, local entrepreneurs live in their villages, have a stake in the successful use of a new technology, and can disseminate the technology simply through their own success.

Sixth, there is a need to generally improve the knowledge base of the people, especially with respect to sustainable methods of land and water management. Since the extension service has the widest coverage in the country, all agricultural extension workers should receive training which includes modern and traditional practices of resource management.

Conclusion

The people of Msanzi learned how to successfully use the water resources of the village, and how to deal with the constraints on full utilization of these resources. Although assistance had been requested from an external agency, the initiative, definition of the problem, and implementation, remained with the village. This form of partnership can be used as an entry point for external intervention in various areas of the environment, for example, control of soil erosion, afforestation, and flood control.

This is an abridged version of a more detailed paper listed in further reading under the *From the Ground–up* series.

Rehabilitation and Construction of Earth Dams in Swaziland

FUNEKILE MDLULI

Introduction

In the period after the Rural Development Areas Programme the Ministry of Agriculture and Co-operatives (MOAC) has come to recognize that to achieve sustained development it is essential to base programme objectives on the real needs and aspirations of target populations. This approach requires more interaction with the rural communities than in earlier programmes. Thus MOAC has developed a so-called indicative agricultural development strategy. The Earth Dams Rehabilitation and Construction Programme is based on the rationale of the indicative strategy. It will give MOAC the opportunity to develop the procedures and techniques needed for the successful involvement of communities in planning and realizing their economic and social development.

During the past 50 years a large number of earth dams were constructed throughout the low rainfall areas of Swaziland. The objectives were to provide a catalyst for economic development and to persuade the communities to remain in their home areas by the improvement of basic amenities.

Although the level of community involvement was initially very low, these facilities became a valuable part of rural life as sources of domestic and stock water, or for cultivating vegetable crops during the dry season. Due to poor design, construction, supervision, lack of community involvement, and conservation of catchment areas, many of the dams fell into disuse through siltation or breaching. There has been growing pressure from rural communities for the repair of these dams. MOAC responded to this pressure, as it provides an opportunity for engaging the affected communities in the sort of dialogue necessary for broader development plans.

The target area

The target area is in the lowlands and covers approximately 400km^2. In this area there exists a large and well-established dams infrastructure. Over time the local communities had come to rely on the dams as sources of domestic and agricultural water. Other developments used the dam water, such as irrigation schemes, fish hatcheries and intensive fish farming, and vegetable gardens, all of which were abandoned when the dams were destroyed in the cyclone of 1984.

The project

By repairing earth dams in the project area, the government hopes to increase economic opportunities through the provision of water. It aims to revive the communities' agricultural activities, to develop appropriate means of involving the local communities in planning their own future social and economic progress, and to introduce the basic concept of environmental conservation and management through a rational land use plan. Secondary benefits are expected to follow from repairing the dams.

Social benefits

The project will have an important impact for women through the reduction in the time and drudgery of collecting water from distant locations. This time and labour could be diverted by women to fruit and vegetable growing. There is also a reduction in ill-health from water-borne diseases.

Economic benefits
These are the revival of previously well-established and successful smallholder irrigation schemes and the reduction in cattle mortalities due to tick-borne diseases. The area is well-served by livestock dipping tanks which have ceased working for lack of water. The consequence has been a higher than normal herd mortality, and a greater susceptibility to tick-borne diseases.

Environmental benefits
These include a reversal of the deterioration of the environment in the programme area, in particular the arresting of soil and vegetation losses through erosion. Also, improved rainfall infiltration and ground water recharge through the construction of conservation structures and reduced runoff.

Implementation
The Land Use Planning and the Economic Planning and Analysis sections of MOAC implement the project. The Land Use Planning section is responsible for the dams, while Economic Planning collects baseline data, and monitors and evaluates the impact of the programme. The organization and mobilization of the local population is a joint venture between the two sections, with the sociologist organizing the meetings. This sociologist works with local communities to form committees where none exist, to establish work parties during the construction of the dams and later will work with groups for education on maintenance of dam catchments and management.

It is essential to ensure the community mobilization and active participation in the programme activities. Effective ways and means of organizing and educating the local population are vital. Visits to the area during the feasibility study indicated that the local population is generally highly motivated and anxious to see the programme started as soon as possible. They were very willing to take their part in a common effort to improve the availability of surface water. As a result, organizing and channelling this willingness to an actual work contribution did not prove difficult. Some areas had already started collecting money to hopefully speed up the construction or rehabilitation of their dam. Such enthusiasm has been evident up to now and is encouraged by the government.

Through their representatives, the local people participated in the selection of proposed new dam sites, and in the detailed planning of preliminary work like bush clearing, stone pitching, collection of stones for gabions, and primary fencing. Mobilizing the local population in this manner always begins with the chief of the area. The project area is governed by three different chiefs. All these chiefs have to be contacted, and meetings arranged for discussion of the project, before one can hope to work with their subjects. The chief is the one who instructs his councillors to announce a meeting in the chief's *kraal*. Here he tells his subjects about the project, and what they are expected to do to facilitate its progress. He will also introduce the government personnel to the people and instruct them to be helpful, that is if he agrees with the ideas of the project and believes the area will benefit from it. It is only after this meeting that it is possible for work to begin.

Programme of action for the mobilization of the local population

The population, through their various traditional forums, are encouraged to elect members to form dam committees or similar organization to formalize their support and participation. Such committees are responsible for:

- Liaising with the Ministry of Agriculture staff who are engaged in the technical aspects of the programme and in the development of a broader programme for the area
- Mobilizing unemployed men and women to be employed as labourers by the contractors building the dams
- Forming rural committees for activities such as removing trees from dam embankments or clearing of sites for dams to be constructed
- Organizing seminars and meetings about the organization and implementation of dam protection and maintenance measures, and erosion control measures in the catchment area
- Monitoring the incidence of water-borne diseases, and
- Collecting money from local population for purchasing fencing for the protection for the dam and catchment.

Through these meetings, the continuous monitoring and evaluating of the project is carried out. Alternatives, such as dam sites or operational methods are reviewed.

Summary

The project has the immediate objective of repairing damaged earth dams in part of the country. This was a need recognized by the people of the area, and is expected to give knock-on benefits from the improved water supplies. Another important aspect is that the project provides the government, and particularly MOAC, with the opportunity to gain experience of working with the people, something quite new in Swaziland. This experience can be applied to the much bigger issue of a national strategy for rural development.

Land Husbandry
Case Studies

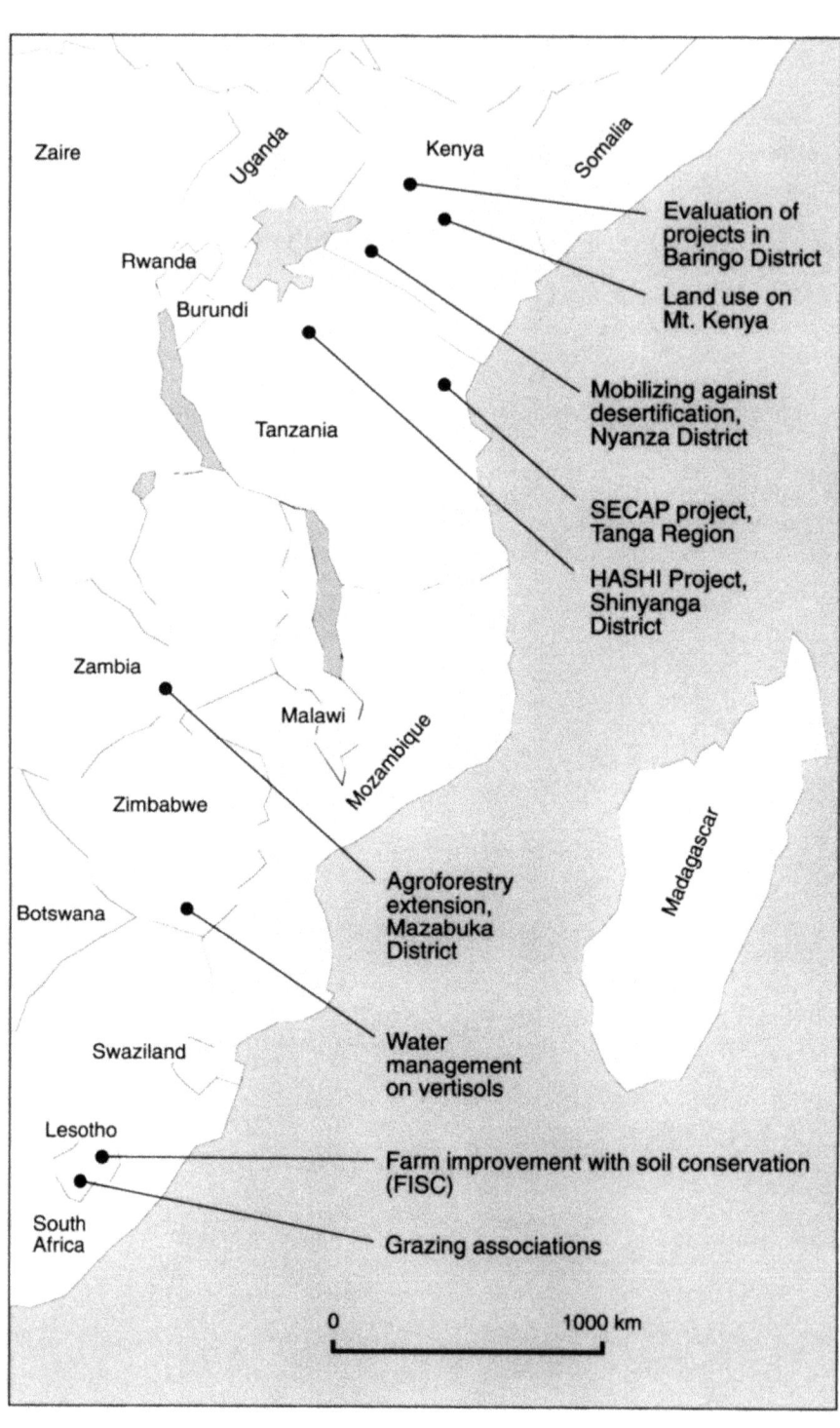

Introduction

N.W. HUDSON

The case studies in this group all focus on particular site-specific problems, such as a different soil with an arid climate, degradation through overuse, politically-motivated reallocation of land, and migration to a new environment. Each approach is appropriate in the right place, but is not being recommended for general application.

The conditions are hugely variable. We have political systems from feudal monarchies to socialist dictatorships, project supporters from tiny local NGOs to national governments and international aid agencies. The farming systems vary widely. But the case studies are all based on grassroots experience of working with farmers, and the experiences are worth recording and passing on.

The papers here are mainly concerned with the rehabilitation or correction of an existing problem. We should also keep in mind the strategy of trying to prevent the problems before they occur. Some examples of positive thinking on this line are in the section about opportunities and benefits.

Several of these case studies include a component of raising the capability of local management. This may be weak for different reasons, such as a history of autocratic top-down leadership, or traditional male dominance, or a restrictive political system. Whatever the cause, improving the ability of local institutions to manage local affairs is high on the list of priorities for better bottom-up development. That issue is examined in more detail in the section with case studies about community development.

A Programme for Farm Improvement with Soil Conservation in Lesotho (FISC)

GEDION SHONE

Background

In Lesotho the lowest administrative unit is the area of a gazetted chief, varying in size from 1000 to 7000ha. Within the chief's area, there are subdivision of villages with village headmen. Higher up in the hierarchy, the gazetted village chiefs fall under senior area chiefs, who in turn come under principal chiefs, who owe their allegiance to the paramount chief, the King. Membership of a chieftain area is commonly a birthright or through marital relationship, or in some cases through application for membership. It is illegal to have more than one membership.

All land in Lesotho is vested on the paramount chief, the King, and gazetted village chiefs are entrusted to administer it on his behalf. Every village chief has a Land Allocation Committee of which he is the chairperson, but owing to the shortage of arable land, the function of this committee is more of land administration than allocation of land to the landless. According to the law, land that has not been planted to crop for three or more years could be revoked and allocated to others. Land can be inherited through the male descendants.

Except for residential areas, all land that is not cultivated is customarily considered as communal grazing land, which does not mean collective ownership but open access for members. This is particularly true of the hillsides and gullies which occupy a large proportion of the arable land. Customarily, some portion of the hillsides is set aside by the chief for the production of thatching grass. Members of the community can apply to the chief to be allocated some portion of the hillside. It is a common practice to give a third of the harvest to the chief in the form of gratuity. The same is true for getting permission to harvest trees that are not privately-owned and fall under the jurisdiction of the chief.

The project

FISC started in 1985 as a pilot project operating in two chieftainship areas. The objective of the project is to conserve agricultural land resources through improved production, soil conservation, rangeland improvement, afforestation, promotion of fruit trees, fodder production, and spring development. The emphasis is on the mobilization of local resources, approaching community groups in the administrative area of a gazetted village chief. The project attempts to introduce simple techniques that are within reach of the farming community, and tries to achieve soil conservation within a system of improving production and not as an end in itself.

The pilot phase was a period of learning. Though there were many changes in approach, the mode of operating could be summarized as participatory development. The pilot phase gave way to phase II in July 1989. At present, FISC covers all the 16 manned extension areas within Mohale's Hoek District – those that do not fall under other donor-financed projects. The project was one of the 10 districts in the country to become an integrated programme under the District Agricultural Officer (DAO). The objective of FISC remains the same as during the pilot phase, that is to optimize benefit from land and land-related resources while sustaining productivity.

Staffing strategy

Apart from two expatriate staff and national experts, the programme uses the services of the staff of the Ministry of Agriculture within the district. In addition to the government staff, the programme mobilizes human resources from among the beneficiaries.

The field staff are the extension, nutrition, and conservation personnel. The first problem was how to use them as multi-purpose staff. While they all have similar background training in agriculture, they are designated with a specialized sector and their allegiance is to the departments and division representatives at the district level. With the consent of the DAO and in line with the government policy of decentralization, these staff members are given training to work as multi-purpose agents. The trainng includes:

- Simple surveying techniques – use of line level
- Map reading and orientation
- Surface hydrology
- Erosion control measures. Biological conservation through range management improvement, afforestation, improved crop cover, grass and other fodder production
- Physical structures such as waterways, diversion channels, terraces, and
- Crop improvement through demonstration plots, and the distribution of farm inputs as incentive to farmers.

The unit area of operation for the field staff is that of the gazetted village chief. The field staff start focusing on one such area during the first year in order to gain experience and have a closer contact with reality and the farmer's needs. The area of operation for individual field staff will expand year after year to cover the adjoining chieftainships. The maximum may not exceed three to four such units depending on the size.

At the level of the community, the Village Development Committees (VDC) and Land Allocation Committees (LAC) are the managers and administrators of communal lands and the human resources within that chieftainship. Training is provided to the VDC members on subjects related to administration and management, including going through laws and regulations governing land and its use and Government Order Number 9 which stipulates the role and responsibilities of the VDCs. This is carried out in co-operation with the office of the local district secretary.

Lead farmers play a vital role in assisting the field staff and they are the media for the transfer of technology from the district agricultural office to the community. Lead farmers are members of the community popularly elected from each sub-village within the chieftainship area. They are trained on simple surveying techniques, control of soil erosion, and improvement and management of land. They form a core of rural workers through whom more and more knowledge could be transferred from all departments of the ministry to the farming community.

Extension approach

The field staff tackle one chieftainship area during the first year while practising the approach. They start by delineating the boundary of the particular chieftainship using 1:20 000 topographical maps. This is followed by making a land resource inventory, and a simple household socio-economic survey. Equipped with the above information, field staff are not complete strangers to the community and the resources available to it. Through a process of consultation with VDCs and LACs they will draw up a first draft of a village activity plan aimed at solving the problems that have been identified jointly, and the solutions as proposed by the members of the community. The plans are intended to optimize the benefit to the members of the community while maintaining sustainability. Soil conservation or land husbandry is not presented as an end itself but as a means to advance production.

The lead farmers provide services to the community and to individual farmers on request. The activities they undertake include the layout of terraces, and survey works as planned by the field staff. The design of waterways and diversions drains remains the responsibility of the field staff together with the district conservation officer.

Relation between field staff and subject matter specialists

An effort is being made to utilize subject matter specialists (SMS) to assist field staff. Regrettably, the programme has encountered some drawbacks resulting from domination of the central administration. The SMSs tend to have their own programmes, which are usually drawn up by their parent department or division. We believe that implementation would have been more effective if carried out through the field staff rather than by the SMSs directly contacting individual farmers.

There is little collective planning at the district level. This is particularly true when it comes to the design and execution of in-service training for the field staff, supervision of field activities, and preparation of annual work plans.

Implementation

Responsibility for implementation of the programme is shared between the community and the government. The community is expected to mobilize and use human and other resources that are available to it. For example, all resident members of a particular community between the ages of 18 and 65 years registered under a chief make up the human resource. They are expected to contribute labour and know-how for communal activities like spring development, tree planting, development of communal orchards, and other activities that are included in the plan for the benefit of the community.

The government provides technical support in the form of materials that are not available in the area and also technical guidance through its field and district personnel. The activities on communal land include the establishment of communal wood lots, communal orchards, communal vegetable gardens, spring development, range rehabilitation, and the construction of conservation structures. Individuals are encouraged to carry out soil conservation measures and are guided by the lead farmers. Individual farmers are encouraged to use agricultural inputs like fertilizers, seeds, and chemicals. These inputs are made available through the VDCs, who buy them on credit and sell them to members at small profit.

Micro-catchment approach

The programme promotes the development of micro-catchments by encouraging those farming within that given area to work together. The intention is to:

- Improve soil and water erosion control
- Enable the planning and management of a physical unit of land with regard to surface runoff
- Serve as a model and training ground
- Plant fodder grasses along the terraces
- Plant fruit trees on terraces, and
- Use improved seed with fertilizer.

The approach does not mean that other farmers will not be served by the programme, it is only a matter of emphasis.

Gully rehabilitation

The community in each chieftainship is encouraged to identify a gully to be treated and made more productive than it is under present use. This involves the planting of the gully floor with Kikuyu grass (*Pennisetum clandestinum*), the sides of the gully to trees, while a strip along the edge of the gully will be seeded with weeping lovegrass (*Eragrostis curvula*) to counter the invasion of the Kikuyu grass into the fields. This way, the gullies are not only checked from further erosion but also made productive.

Inducement and incentive payment

Payment has been made to farmers who have participated in the construction of structures of a public nature at the rate of half of the government minimum daily wage. Payment is now made only for the construction of diversion canals and waterways, while all other conservation activities are done on a voluntary basis.

The programme also provides incentives in the form of seeds and fertilizers to farmers who have terraced their fields. The intention behind this is to demonstrate the combined effect of physical conservation with grass seed and planting material for fodder grasses, and fruit trees.

Formation of the board of farmers

The sharing of responsibility and the transfer of knowledge to the farming community cannot be achieved without strengthening local institutions. To ensure the fullest participation of the community, a board of farmers has been formed, composed of two representatives from each of the 16 extension areas. One of them is the chief of the area, and the second is an elected member of the VDC. The tasks of the board are:

- To participate in monitoring the services and activities of the programme
- To represent the interests of the programme's clients, the farmers
- To formulate and introduce management tools that govern land resources and manpower resources
- To initiate and manage any activities that will benefit its member communities and act as a common service to them, and
- Most important is the creation of a forum for the full participation of the population in shouldering the responsibility of farm improvement.

An example of on-going activity under the heading of common services is the purchase and distribution of farm inputs. Previously done by the programme, this is now carried out by the community, and is a significant step towards autonomy and self-management.

The board has asked the programme to provide guidance for implementing grazing rights. Another scheme being introduced is that of the village veterinary workers (VVW) or so-called barefoot veterinarians. Candidates from the community are chosen by the VDC. They are trained in the identification and treatment of common diseases. Equipment is supplied to the VVWs through the VDC. They purchase drugs through the Ministry of Agriculture at wholesale prices. The small profit obtained through this practice goes to the individual VVW, while cattle owners benefit by obtaining a locally-based service.

The effort towards people participation

Let us look at the degree of participation of community members in specific activities.

Range land management and gully reclamation

The degradation of the range lands in the country contributes to further damage to the crop land and aggravates the formation of gullies. Gullies and range lands are considered as communal areas and used for grazing. Such land belongs to everybody, but is nobody's responsibility. Those that have livestock are the ones that benefit from it. They maximize the exploitation of such communal resources. The tenure of communal lands can be summarized as collective open access, rather than collectively owned.

Here is a typical example where the cause and the possible solution for soil erosion are directly linked to the social structure. Those who have livestock are interested in maximizing their profit at the expense of land degradation, while the non-livestock owners are indifferent to the quality of the communal land resource. The system of grazing rights addresses the issue of equitable distribution of the range land. All members of the community would have an equal and marketable right, irrespective of livestock ownership. Several meetings were held to explain operation of a marketable grazing right system. Mock exercises have been carried out on the procedure of transfer of the rights. The board of farmers have agreed to adopt this system as a tool to properly manage and improve the range lands. If adopted, it would invite all members of the community to participate in the improvement of the range land. This result would benefit everybody, whether owners of livestock or not.

The adoption of this method has been discussed with the community at various levels. Livestock owners and most community leaders were opposed to the idea as they are used to free access. But with the recent intentions of the government, which is determined to introduce a grazing fee as a means of controlling range land degradation, many have resorted to adopting the grazing right system.

On the question of gully rehabilitation, discussions were held with leaders of the community, opinion leaders, and the members at large, in an effort to involve all members of the community in the rehabilitation effort. In some areas the decision was reached to allocate the gullies to the landless. In such cases, the new owners would be required to develop it for fuelwood, fruit trees, fodder production or a combination of these and other ventures.

Spring development
Spring development as a source of domestic water supply is a popular practice. Because of the land tenure, springs that exist on communal land are not fully utilized. This is one of the activities that is included in the communal resource development activity plans. Combined effort by the community and government is made to harness this resource for collective use. The water is used for developing communal orchards or vegetable gardens or for a village nursery.

The experience of managing communal enterprise is unfortunately not encouraging. Many are frustrated to the extent that it is difficult to initiate activities of this nature. The solution has been to privatize the use of such schemes through lease systems. The plots are auctioned at the beginning of each season. In theory, landless members should get precedence in the allocation of plots.

Crop land management
Most of the cultivated fields in Lesotho have been terraced since the beginning of the century. Maintenance and the construction of cut-off drains and waterways are the major physical conservation tasks. The predominant task is improving the fertility of the land through better management practice.

Research in the field of crop production is far from satisfactory. The major crops are maize and sorghum. The recommendation is focused on the use of fertilizer and improved seed. The research findings are from more favourable places as compared to the drought-prone district of Mohalae's Hoek. Because of the drought risk factor and the ever escalating price, farmers are using only thirty to fifty per cent of the recommended rate of fertilizer and still obtain a significant result in good years. The research is focused on the use on inputs and not on crop husbandry like composting, and intercropping which would be within the reach of most of the farmers.

The number of landless people has been increasing over the years. The present estimate is close to thirty per cent. On the other hand, many of the landowners are absentee farmers, restricting a timely farm management decision-making process. Over ten per cent of the landowners are widows. In this situation, the programme has resorted to introducing labour-saving services such as establishing contour grass strips.

Of the total arable land, close to twenty per cent remains fallow annually. Though fallowing is recommended from a land management point of view, it reduces grain food production, and is equated to misuse of land. It can lead to the revocation of the tenure if not cropped for more than three consecutive years.

One proposition suggested to the board of farmers has been to initiate the VDCs to act as middlemen and guarantors between the landless or other interested individuals, and owners of fallow lands. In the case of land that is held by widows or families who lack the capacity to manage the land for optimum use, it is recommended that the land is leased for a longer period of time. It is also advised that the lease holder cultivate perennial crops such as fruit along terrace edges.

Incorporating the FISC programme into the extension service

National seminars have been held for the 10 district agricultural officers and the district extension officer. The approach is slowly gaining support. Responding to this acceptance, a small unit is now attached to the Department of Conservation, Forestry, and Land Use Planning. The main task of the unit is to disseminate the approach. It will also identify and incorporate activities that have gained popularity and acceptance by the majority of the farming community. These will then be spread through training. The training programme is aimed at the training of trainers. This starts at national headquarters, then the SMSs at the districts, who in turn will be training the field staff of their districts, and the field staff will be training selected farmers within their extension areas, thus having a multiplier effect.

Mobilizing Farmers to Counter Desertification in Nyanza District, Kenya

SR. DOLORES RAUCH

Introduction

1984 marked the climax of a four-year drought in western Kenya. Among the most seriously affected areas was south Nyanza District which borders the eastern shores of Lake Victoria, just south of the equator. Aid for the affected population arrived from various sources, including the Catholic diocese of Kisii.

Aside from immediate stop-gap measures in terms of food and other necessities, far-sighted diocesan leaders began researching the root causes for the calamity. Scouting the barren hillsides and deeply eroded gullies, they realized that resource degradation had reached frightening dimensions. 'Replant those trees, or south Nyanza will turn into desert within five years', warned Chris Pable, the European water expert who had been called upon to help them analyse the situation. The diocese resolved to meet that challenge through concerted local efforts. Mobilizing against desertification (MAD) was born.

Casting about for a suitable model and finding no clear direction in current literature, the planning team decided to devise a research design based upon their own observations in south Nyanza, including the capacities of the land, the characteristics of the people, and cultural traditions. The Ford Foundation of the United States promised to support the venture.

The research design

Defining the target area
The planning team for MAD began its work by defining the target area for the proposed efforts to stem advancing desertification. They decided to attack the threatening menace at its source, a 2500km narrow strip of land bordering the eastern shores of Lake Victoria which has become identified as low-potential south Nyanza.

Capabilities of the land
From an ecological point of view, the target area presents a complex picture, featuring five agro-ecological zones. MAD's agricultural endeavours have been most successful in LM3 and UM4 zones.

Agro-ecological zone boundaries closely resemble the patterns of isohyets, but the zones actually contain a series of differentiated agro-ecological microcosms. Soils differ greatly in fertility, depth, and drainage. Most of them have been strongly affected by the forces and desertification, especially erosion caused by deforestation and overgrazing. However, water is the determining factor for successful farming, with variations of soils playing a secondary role.

Characteristics of the people and cultural traditions
Four hundred years ago, Luos, a Nilotic people originating from the valley of the White Nile, pushed their way south and east to take possession of the land bordering the eastern shores of Lake Victoria. Traditionally nomadic pastoralists, they gradually settled in the forested uplands and the lush savannas, grazing their herds of indigenous sheep, goats, donkeys, and cattle. Subsistence farming to produce the family's food supply was

left to the women who laboured with hoe and panga to wrest from the soil crops of millet, sweet potatoes, and cassava.

Farming is still carried out in this fashion. Isolation forced upon the people by poor, often impassable, roads, by physical poverty and widespread illiteracy (sixty per cent) has left them behind their more favoured neighbours in technical and economic development. Having only recently come into possession of private property (land demarcation started in 1987 and continues in some areas until now), Luos also lack a traditional farmer's attachment to the soil, a value system derived from a sense of freedom and personal initiative to develop private means of production.

Attitudes and mores derived from the cultural heritage left Luos unprepared for the socio-economic changes of recent decades. Population increases and mounting demands for ready cash forced them to make ever greater demands upon the resources of this fragile environment, especially its natural vegetation. Landowners now reap the consequences of a man-made ecological disaster. Ignorant of the cause and effect relationships inherent in their dilemma, they react with bewilderment and a sense of hopelessness.

To reverse the process of desertification, the planning committee for MAD decided upon a number of basic principles and objectives.

Principles

These are:

- MAD targets the most deeply affected area which is low-potential south Nyanza
- A bottom-up approach is followed. It implies strong involvement of the local people through sharing of past experiences, and of visions for the future. As well as participation in planning and concerted efforts to achieve objectives
- MAD begins its work with established local church groups, and spreads outward to the total population
- Local personnel receive long-term training to enable them to spearhead and manage the programme, and
- Careful observation and adaptation to developing insights into local needs and resources will give direction to the project – not a definitive, imposed blueprint.

Objectives

Raising ecological awareness among all levels of the population
The message to the religiously inclined, but fatalistic, population must be that the problem is man-made, not an act of God. We reach together for ways and means to reverse the process of desertification.

Establishment of tree nurseries
Raising the level of understanding and motivation leads to increased demand for seedlings ready for planting. In response to that expected

demand, MAD established one tree nursery in each of the nine Catholic parishes located in the target area. They serve as models, while providing readily available seedlings for the local population. Training of nursery workers takes place on-site, as well as through well-designed workshops.

Planting of trees and caring for them
Though much of their land had formerly been heavily forested, Luos never practised silviculture. Used sparingly in their subsistence economy, trees regenerated naturally. The indigenous population did not grasp the economic value of trees until encroaching westernization imposed upon them the need for adopting a cash economy. They chanced upon trees as a ready source of income in energy-poor Kenya. Not comprehending the importance of trees in tropical ecosystems, they were unable to foresee the dire consequences of their actions.

MAD leaders recognized these problems, and their causes. Counting on the strong religious beliefs of the local people, they devised a simple theology of ecology based upon the creation story. Parish priests cooperated by incorporating scientific information within the religious context of Sunday worship. MAD personnel were at hand to demonstrate techniques for planting the five seedlings allotted to each family during the first planting season. People eagerly participated in the fight against desertification, the enemy they had come to recognize and fear.

A bottom-up approach to development work
To become truly operative, this concept must permeate all relationships among administrators, staff, and local participants. In short, it must become embedded in a teamwork management style.

True to the basic tenets of the chosen model, local persons are selected for staff positions. Since most of the privileged few who are able to earn the academic credentials required for further study and lucrative employment regularly leave their home territory, the local pool of potential employees consists of the remnant of the original population. Native intelligence, interest in trees, and dedication to their home area, became the major criteria for staff selection, rather than earned academic credentials.

Such recruitment practices require heavy emphasis upon staff development. It is an on-going process and takes place informally through positive normal interaction among the MAD team. It is supported by policies, forms, and structures which are outgrowths of past experience and felt needs, rather than artificially injected techniques. Staff meetings provide opportunities for more formal training. Team members take pride in chairing meetings and taking minutes. They bring concerns and questions to bi-weekly meetings and assist each other with finding solutions. They learn to accept and render constructive criticism and to engage in desirable forms of conflict resolution.

Growth in personal maturity and professional competence requires opportunities for experiences designed to deepen insights and broaden intellectual horizons. Planned field trips to programmes with objectives similar to MAD's, as well as workshops conducted by external resource personnel, serve that purpose, as do interactions with consultants. The recent establishment of a small staff library also makes significant contributions to staff improvement. Opportunities for formal training courses are presently under consideration.

Rationale for change
The original design for the programme did not contain the 'model farm' concept. Model farms evolved gradually from insights into the complexity of the causes of desertification, and the necessity for adapting land-use practices, to the present agro-ecological situation of low-potential south Nyanza.

As MAD staff interacted with the local people, they gradually began to understand the importance of working closely with individual land holders, recognizing them as the true custodians of the land. Only they can ensure the survival of newly-planted trees on their private property, and the success of reforestation programmes. They hold the key to a healthy ecosystem. So, the MAD team set about initiating a suitable programme for interested farmers.

Objectives on model farms
These are:

- To learn more about trees and their survival under various conditions
- To cultivate a loving relationship between farmers and their land
- To develop among farmers an understanding of the land as a means for production and economic wealth
- To teach farmers the skills needed for achieving a system of sustainable agriculture
- To assist farmers with reclaiming degraded land and improving soils through good land management practices, including agroforestry, organic farming, erosion control, and water management
- To enable farmers to appraise the capabilities of their land, and devise production systems in harmony with it
- To support farmers in their efforts through moral and financial assistance, and
- To encourage farmers to serve as models to their neighbours through sharing of experiences and rendering mutual support.

Methodology
The tutorial stage The model farm programme originated with four one-acre plots, set aside by selected farmers for experimentation with agroforestry systems. The chosen farmers had attended MAD awareness workshops and had been recommended by their local communities on the basis of two criteria, being hard working, and being open to change.

With their appointments began an intensive tutoring relationship between staff members and individual farmers. Since MAD staff were lacking in some of the skills related to farm planning as well as to water and soil management, local agricultural extension workers became willing and able partners in this endeavour. Two workshops conducted by KIOF gave additional input on organic farming techniques.

Initial efforts to clear useless brush from wasteland met with jeers and dire predictions on the part of neighbours. However, after they witnessed the first crops, once-sceptical farmers asked to become participants in the programme. The time had come for expansion.

Workshop/extension stage MAD organized its first farmers' workshop in May 1990 and invited model farmers, whose number had grown to 16, to bring with them interested neighbours. Most of the 30 visitors requested membership in the programme at the end of the workshop. The list of active participants grew to 40.

Now, an intensive second phase for the model farm programme began, consisting of a sequence of six workshops interspersed with extension services. Workshop presenters continued placing emphasis upon deepening environmental awareness, while providing skill-building exercises. All sessions concluded with the drawing up of action plans covering the period between workshops to foster accountability. Skills taught at workshops included aspects of agroforestry, contouring with simple A-frames, soil and water management practices, organic farming methods, record keeping and accounting techniques, and grafting and budding of fruit trees. All are geared to achieving the ultimate aim, namely to develop systems of sustainable agriculture that will counter resource degradation and desertification.

Cluster farmers extension model stage The model farm idea is spreading rapidly. More and more neighbours desire to join the programme. Some of the model farmers spontaneously began assisting them. They have now become the nucleus for the MAD cluster farmer extension model. The project was launched during a workshop in April 1991 when 12 enthusiastic model farmers, serving as organizers and leaders, met with MAD staff to discuss basic principles, expectations, and relationships involved in the new venture. It is hoped that the extended model will lead to closer interrelationships among farmers and communities, and effective co-operation for solving common problems.

Conclusions

The past several years have been tremendous learning experiences for all MAD participants. Originating as a straightforward tree planting exercise involving local parish groups, MAD has developed into a complex force, shattering narrow horizons and freeing new energies within an environment characterized by widespread ignorance, poverty, and hopelessness. It channels the new energies in to the will to build a better socio-economic as well as physical environment. During the short span of four years, MAD has succeeded in establishing 40 small examples of potentially sustainable agricultural systems within degraded UM4 and LM3 agro-ecological zones of low-potential south Nyanza, on land which has lain idle.

The success of the model farm programme, especially in eliciting ready acceptance of innovation and the rapid dissemination of new technologies, indicates the basic soundness of the MAD model for development. Changes have been brought about through modest financial inputs, coupled with innovative approaches to research and development. Close attention has been paid to water and soil management.

All genuine development must imply self-development. The energies required for this process can be released through a bottom-up team model which releases the potential of individual participation through supportive mutual internal relationships, and strong external support linkages. Members of the MAD team claim that their experience has challenged and enriched them, which significantly contributed to the success of the programme.

Other insights seem worthy of note

- Sustainability of development efforts cannot be achieved unless sustainability is the main focus for a programme from its very beginning
- No programme can successfully work in isolation. Staff and participants need to consciously foster linkages within their milieu, including NGOs and government extension offices
- A participatory team model facilitates innovation and diffusion of technological adaptation outside the core area
- Socio-economic and technical development must mutually support each other. There can be no progress toward holistic development in an atmosphere of competitiveness with or isolation from each other
- Changes, whether physical, socio-economic, or cultural, are often painful to human nature. Innovation becomes less threatening if someone walks with and supports the affected person, as do MAD supervisors through regular support visits to model farms
- Flexibility on the part of funding agencies makes innovation possible
- Careful planning and accountability are vital to success
- Tree planting exercises alone cannot bring about permanent solutions to an impaired environment. Farmers ultimately hold the key to successful counter-measures to desertification, and
- MAD has not yet found solutions to all problems prevalent in the target area. Shortages of water pose the most serious threat to agriculture in South Nyanza. The situation is under investigation. Another situation which needs careful research and effective improvements is the lack of good roads. Poor roads inhibit communication and prohibit the establishment of adequate marketing systems.

In brief, good beginnings have been made in south Nyanza. However, there is much work to be done for creative research and development efforts.

Farmer Adoption of Improved Water Management on Vertisols in Semi-arid South-east Zimbabwe

P. NYAMUDEZA, E. MAZHANGARA, T. BUSANGAVANYE, and E. JONES

Introduction

Ndowoyo communal area in the south-east of Zimbabwe is a gently rolling plain at an average altitude of 430m asl. The soils are mainly vertisols. The rain season, October to April, is warm to very hot with an average rainfall of 570mm. The long-term average rainfall for October to February, the critical period for crop growth, is 490mm, ranging from 223 to 723mm with an extremely erratic distribution. During the rains the mean maximum monthly temperature ranges from 31 to 34°C, with absolute maximums of 40°C and above. The mean minimum monthly temperatures fluctuate around 20°C and the open-pan evaporation is about 1100mm. The winters are generally dry and cool.

Agricultural background

Ndowoyo is in Natural Region V, an area unsuited for any form of arable cropping (Vincent and Thomas, 1960). The main farming activity is rainfed cropping combined with cattle herding, with sorghum as the major traditional crop. Maize is also grown extensively, and cotton and sunflowers as cash crops are on the increase. If the rains falling before mid-November produce a flush of weed growth, crops are sown. Thereafter dry sowing is practised in expectations of rains, but these can be delayed until the end of January. Farmers grow improved sorghum and maize varieties with short duration alongside the long duration traditional varieties. Farmers use no cattle manure or inorganic fertilizers because 'they burn the crops'.

Crop performance can be summarized as follows. In five years sorghum and cotton produce one good crop, three mediocre, and one failure, maize produces one good crop, two mediocre, and two failures.

Constraints on crop production as viewed by research in 1982

The Lowveld research stations in 1982–83 embarked on a programme to improve the production of crops grown under rainfed conditions. Consideration of historical rainfall amounts and distribution over a 28-year period showed that in more than half the seasons there was inadequate rain to grow a full crop (that is a crop with the plant density recommended for conditions of adequate rainfall). Mid-season droughts occur frequently and can cause yield reduction or crop failure even in years with adequate total rainfall. Hussein (1987) in a detailed study of growing seasons in the semi-arid areas of Zimbabwe concluded that the probability of having a season length of 110 days, the minimum for most crops, was 16 to 23 per cent, and a serious mid-season drought occurred once in four seasons. Once in 19 years or less there was no season.

Research programme – the theory

The two aims of the rainfed crop programme were, first to try and grow less than a full crop, on the assumption that it would require less water. Second, to develop techniques that would increase the water available to the crop.

In this context something less than a full crop is a crop sown at a lower population than the usual recommendation. Fewer plants for each unit area of land would benefit the crop by reducing the leaf area at all stages of growth, and such a crop would transpire less water for each unit land area per day, and therefore survive a longer period between rainstorms.

Rainwater is lost to crops in two ways – losses from runoff during heavy rainstorms, and evaporation of water from the surface of the soil. Preventing the loss of water by runoff during storms can be achieved by increasing the surface storage. The most efficient method of achieving this is the creation of ridges and furrows with the furrows blocked at intervals. This system is known as tied ridges or diked furrows.

Another advantage of tied ridges is that a portion of the rain runs off the steep sides of the ridges and concentrates in the furrow, where it percolates deep into the soil and is protected from loss by evaporation. This concentration of water in the furrows in the early season made it sensible to sow the crops in the furrows, a practice that led to the term tied furrows, rather than the more usual tied ridges.

Further advantages of the tied furrows are that by holding most of the water on the land where it falls, soil erosion is virtually eliminated, and waterlogged patches do not develop in any natural hollows. In wet seasons waterlogging is a problem in the hollows of the gilgai micro-relief, a natural feature of vertisols.

Research programme – the results

These theoretical considerations led to trials on the Chisumbanje Experiment Station in 1982–83, and on the station and sites on farmers' fields from 1983 onwards. In the first two seasons flat-bottom basins were tried, but in subsequent years these gave way to the more practical tied furrows. In addition to testing methods of preventing the loss of water, the trials also tested the three main crops, sorghum, maize, and cotton, with a wide range of plant populations less than a full crop. The results are summarized below.

Effect of growing less than a full crop
Low, medium, and high populations of sorghum and cotton produce very similar yields in good and average seasons, because of the ability of these

crops to compensate through tillering or producing larger frames. In poor seasons or seasons with a serious mid-season drought the lowest populations do have some advantage. Maize cannot compensate for lower populations and yields are reduced in the good seasons, but in the poorer seasons the lower populations tended to produce some yield when the higher populations failed completely. The current view for all three crops is to sow half to two-thirds of the full crop population because the possibility of a gain in the poor seasons is more important than an increased yield in the less frequent good years.

Effect of growing crops in furrows
Over eight seasons of trials the average advantage of sowing in furrows over sowing on the flat was, sorghum twenty-five per cent, maize forty-six per cent, and cotton thirty-two per cent (Jones *et al.*, 1989). Yield levels of all three crops varied widely from season to season. However, sowing in furrows tended to produce heavier yields in all seasons. The smallest increase in yield, less than five per cent, was in the one good rainfall season, 1984–85, when sorghum produced 3.7t/ha, maize 2.7t/ha, and cotton 2.5t/ha. The largest increase from growing in furrows more than doubled the yield from the flat for all three crops. These were recorded in seasons with average total rainfall, and did not occur in the same season for each crop. The increases in yield in the driest seasons ranged from about ten per cent to more than one hundred per cent depending on the severity of the mid-season drought. In most seasons the difference in yield was apparent in a visual difference in growth of the crops, especially during the course of a mid-season drought when furrow crops suffered less wilting, or showed wilting symptoms much later than those on the flat. The crops in 1987–88 were exposed to the heaviest two weeks of rainfall on record, 22mm in 14 days, one week after germination. During the rain the crops in the furrow suffered more than those on the flat. But they recovered and survived the following mid-season drought better than those on the flat and produced twelve per cent more sorghum, one hundred and thirty per cent more maize, and forty per cent more cotton.

The results summarized above are for furrows 1.5m apart. Two other furrow sizes were tested, 1 and 2m. In a good season the 1m furrows tended to give the highest yield, whereas in a poor season the 2m furrows tended to be best. But over a number of years, the 1.5m furrows were the more consistent. In the fifth season the 2m furrow was discarded because of the problem of construction and maintenance. The 1.5m tied furrows survive the beating action of the rain better than the 1m, and therefore are capable of holding more water through the season. Furthermore, in the 1.5m tied furrow, land preparation between crops consists of ripping or ploughing out the line of the old crop in the furrow. This is followed by the rebuilding of the ridge. The 1m furrow requires full ploughing, and complete rebuilding of the ridges in most years. The 1.5m tied furrows produce permanent ridges, which enable the farmer to accurately set the alignment across a slope by minor annual adjustments. The ridges also help to maintain this alignment in future years. For these reasons the 1.5m furrow is the preferred width on the vertisols.

Farmers' exposure

The farmers in Ndowoyo have been exposed to the furrow technique in the following ways:

- The seven trial sites on farmers' fields have allowed farmers to watch developments at all stages during the season
- On three occasions in the second year small informal field days were run on the farm sites for the benefit of the farmer and his neighbours
- Every year from 1984 to 1989 the experiment station has organized a formal field day near the time of harvest. Farmers from up to 50km away were brought in and a senior government or extension officer attended. The number of farmers attending, men and women, ranged from 200 to 500 and local school children and teachers made the attendance over 800 in recent years, and
- The staff of the experiment station are related to the local farmers or are farmers themselves in their spare time.

Farmers' reactions

The reaction of farmers and the extension workers in the early years was negative. The reaction to growing less than the full crop was that it is a waste of land. The reaction to tied furrows was the acceptance of the principle of holding the water. But there was doubt that they, the farmers, could construct these with the labour force, draught power, and equipment available to them. Waterlogging in wet years was also a concern.

This attitude continued for the first five years, but attendance and interest in field days increased. Every season they saw that the crops in the furrows were as good as or better than the flat, and in the wet seasons, 1987–88, their fears of waterlogging proved groundless.

In the sixth season a few farmers asked for furrows and these were constructed with a station tractor. This generated a lot of interest, and in the seventh season the government tractor hire service, responding to farmer pressure, declared a price for constructing furrows, although they failed to supply a ridger. Late in the season a ridger was obtained from a local sugar and cotton company, and a few farmers were able to have their land furrowed at a commercial price using privately-owned tractors. This represented a major change in attitude of farmers, who only two years earlier considered furrowing to be impractical. Now they were willing to pay cash for the work.

In the eighth season, 1990–91, the sugar company, following the attendance of their management at a field day offered to furrow 100ha of land with a tractor with a view to increasing cotton production for their ginnery. The 100ha were allocated to the areas around the trial sites, with each farmer limited to 1ha and asked to pay half the commercial price.

The offer was over-subscribed in cash one month before the operation was due to start.

In 1989 the station modified a standard ox-drawn ridger to enable it to build and maintain 1.5m furrows, and this was demonstrated to farmers on one of their farms. Subsequently, the ridger has been borrowed by many nearby farmers, and farmers groups further afield have asked for their own ridger. The ridger can be pulled by two strong oxen or four average animals.

Two aspects are impressed on farmers who put in furrows. First, the furrows should be aligned across the line of water flow in their fields in order to discourage the flow of water along the furrows. Second, furrows must have cross ties because open furrows without cross ties can lose more water, and cause more erosion than flat land.

At the start of the 1990–91 season the number of farmers who had some of their land in furrows was estimated to be more than 150. One hundred had their land ridged by tractors provided by the cotton company, and the remainder had it done by private tractors, the ox ridger, or simply using their ox ploughs.

Socio-economic aspects

Informal survey
In 1989–90, the eighth season, an informal survey was conducted to assess the farmers' views of the tied furrow technique. The survey comprised 18 monthly visits, and interaction with 20 to 25 farmers. The farmer views on the tied furrows can be summarized as follows. First, the majority of the farmers visited were aware of the tied furrow system of farming and how it helps with the holding of rain water. Second, without exception the farmers would be willing to try the new system on their farms, but had not done so because of the absence of suitable equipment or lack of money to pay for the work.

The formal survey

A formal survey was conducted on 35 farms in April 1991. One objective was to determine the factors affecting the adoption of the tied furrows to date, with a view to guiding future extension strategies. The findings include:

- A positive correlation exists between adoption and research contact, indicating that on-farm trials of an improved system strengthen the possibility of adoption.
- Nine out of 10 farmers indicated a preference for making furrows with a tractor because of the probability of having no oxen, or oxen too weak to work, as a result of the frequent drought. The one in 10 who opted for ox ridging lived near an irrigated estate where farmers cut and carry grass from the canal banks to feed their cattle, and

- The majority of the farmers would be willing to borrow to pay for constructing the tied furrows.

Economic viability

In the first year of ridging farmers incur an extra cost of Z$50 a hectare for hiring a tractor and ridger, and the labour for cross ties. On average, this will be more than compensated by the value of the yield increases and by a reduction in seed and weeding requirements for crops sown in the 1.5m furrows. In most seasons, weeds only grow in the furrows, and farmers with furrows have said that this reduces the weeding load.

In subsequent years farmers will need to rebuild ridges, an operation similar in draught power requirements to ploughing and sowing on the flat. The total cost is therefore the buying of an animal-drawn ridger at the current retail price, about Z$400, this amortized over 10 years at twelve per cent interest and allowing for repairs and maintenance works out at Z$18 a hectare for a five hectare farm. Again, this is considerably less on average than the value of the expected yield increases. The benefits of reduced weeding and less seed remain to offset the labour for making the cross ties.

Summary

The research programme to date has established the following facts:

- Tied furrows increase and stabilize yields of all crops
- Many farmers understand the principles of the tied furrows and they are interested in adopting them
- Tied furrows can be economically viable, and
- Tied furrows were considered impractical by everyone until they had seen their benefits in dry and wet seasons.

Conclusions

The decision on whether a new technique is practical or not should not be taken until information is available on the level of benefits to be derived from its adoption. Also, it is unreasonable to expect farmers to adopt a new technique until they have themselves witnessed its performance under the range of conditions existing in their area.

Acknowledgements

The authors wish to thank the Director of Research and Specialist Services for his continuous support and encouragement, and the International Board for Soil Research and Management (IBSRAM) for its financial support from 1987 to date. The diligent work of E. Kamba and G. Mhlanga in the day-to-day supervision of the trials is much appreciated.

An Agroforestry Extension Project in Mazabuka District, Zambia

YEMBO KAONGA and ELIZABETH MALAYISHA

Introduction

Family Farms is a non-profit-making rural development NGO working in the Southern Province of Zambia. Its main objective is to foster agricultural production in Zambia. The organization is not directly engaged in farming, but renders services to the farmers to achieve its goals. The services are soil and water conservation, borehole drilling, fishing development programmes, and hardware shops.

In 1986, an agroforestry extension project was introduced within the soil conservation programme. It located in Magoye Agricultural Block of Mazabuka District, with a population of 27 000 small-scale commercial farmers. There are two land tenure systems in the project area. Traditional land is allocated by the chief or village headman, and traditional laws are applied. There is also state land which is parcelled and allocated to farming families. Here farmers have leases or title deeds.

Objectives

The main objective of the project is to assist farmers to increase agricultural production through the use of sustainable practices in crop production, pastoral farming, and forest farming, without causing adverse effects to the environment. The second objective is to extend income-generating activities beyond the single cropping season. This is done by encouraging the on-farm production of trees and tree products.

Farmers' problems

A needs assessment survey was conducted in the project area in the second year of the project. The survey revealed that the needs were different in the two land tenure systems. The reason is that traditional land has a long history and is densely populated compared to the settlement schemes, which are only about 20 years old. The study revealed that very few farmers have their fields on sloping land and that soil erosion by water is not a serious problem. However, wind erosion is a serious problem. Over-stocking, over-grazing, and shortage of fodder during the dry season, are worrying. In reserve land, fuelwood and building poles are in short supply. For most of the farmers, live fences are a great need. The study also looked at the experience farmers have with certain trees either as fruit trees or manure trees, and whether the trees have been planted by the farmers.

It is evident that land degradation is of concern in the area. Farmers have complained of reduced yields from their fields. The colour of water in rivers during the rainy season suggests that erosion is more serious than is shown in the surveys. Farmers have also complained about the changing rainfall patterns.

Solutions

The project addressed itself through the following issues.

How to farm with less fertilizer

The project does not tell the farmers to do away with chemical fertilizers, but that they should consider alternative cheap means of improving production, for example growing of manure trees like *Acacia albida* in their fields, green manuring, and application of kraal manure, which can be composted with crop residues.

How to reduce soil erosion on sloping fields

We discuss with the farmers the importance to plough along the contour, and how to find a contour with the use of the A-Frame. We encourage planting of trees or shrubs, or a combination, on the contours. These help to mark the contour permanently in the field. These need not be nitrogen-fixing trees, but according to the farmer's preference.

Mechanical means of soil erosion control

This is addressed by storm-drains, waterways, ditches, and graded contours.

Windbreaks

Farmers are reluctant to plant trees as windbreaks across fields, unless it is a farm road or farm boundary.

Growing trees for income generation

This includes growing fruit trees, fuelwood, and timber for building, trees for fodder, and live fences. Advice is given on raising seedlings.

The soil conservation measures are not new to the farmers. Their background during the colonial era has a lot of impact on their adoptability by farmers, as some were enforced on them to the extent of paying fines.

At independence, highly subsidized fertilizer meant that all labour-intensive compost-making and green manuring were abandoned. The official recommendations during this time entailed clearing all trees in the field and did not favour trees like *Acacia albida*. A farmer with such trees would not get a loan. At present, fertilizer prices are out of reach of poor farmers. Our project looks at whether agroforestry practices could be the answer and how the farmers can incorporate them in their farming practices.

Extension methods

In the project area training and visit (T and V) was the extension method used by the Ministry of Agriculture staff when the project started. We were not able to interact with farmers directly through the camp officers responsible for T and V. Establishing relationships with extension agents and farmers was not easy. We were unknown by camp officers, and they expected to get some sort of an allowance from us if they promoted our messages. The camp officers were unfamiliar with agroforestry. They started to attend our meetings out of curiosity. A relationship with them was established. Subsequently, camp officers became the means to help us identify the contact farmers and to mobilize others. Now camp officers and farmers attend meetings regularly to create awareness and to discuss implementation, adoption, and monitoring and evaluation.

The top-down approach

In the initial stage of our project, we had adopted the top-down approach. We were the body with all the knowledge. No needs assessment study was done. Information was delivered to the farmers as we considered appropriate. We expected a lot. Farmers are reluctant to take up some practices. When that happens they watch fellow farmers who have tried these practices, and learn the problems they faced.

Farmers presented our meetings. There was no monitoring. When we looked, we realized that we knew very little about the needs of the target group of our message. This marked the end of the top-down approach. The birth of the bottom-up approach came in the second year of the project.

The bottom-up approach

In the second year our approach changed. A detailed needs assessment study was carried out. The results were discussed. We appreciated what farmers contributed to meetings. It is here where we learnt of the *Acacia albida* as a useful tree in respect to manuring the fields and providing fodder for animals.

The farmer who favours the *Acacia albida* tree makes comments like: 'I clothe and feed my family under the trees in my fields'; 'my yields are not drastically affected during the drought years'; 'the maize plants under the *Acacia albida* do not wilt even after a dry spell of two weeks'; and 'I do not apply fertilizer under the trees'.

Those against trees in fields give some important points that cannot be overlooked: 'The tree can easily take over your field if you leave it fallow for more than two years'; 'It is thorny'.

There was a complete participation with the farmers in most cases during the discussions. Our role was to introduce the subject to the farmers and to summarize the outcome of the discussed issues. A farmer would volunteer to register with us to take up a certain agroforestry practice that pleases him or her.

The camp officer who is well-known by farmers plays an important role by leading the discussions. The camp officer is drilled on all topics we offer to the farmers.

Drama for development

This is a recent extension approach which we introduced, and has worked quite well. The selected topics and the information disseminated was clearly noted by the farmers. In one sketch, we tried to show the importance that at all times, the solution should be reached by both wife and husband in the family. We showed how decisions made by one person in the family can affect the others. For example, the problem of fuelwood is much felt by the wife, whereas the husband does not really see it as priority. He is more worried about the weed infestation in the field.

Visits, demonstration plots, field days

Individual farm visits to the farmers are made in three stages. First and second visits are made in January to deliver trees and check whether or where the trees were planted. The third visit is done in May towards the dry weather. Farmers are advised on how to protect the plants from fire and animals. The fourth visit is in August. This is our planning visit when we sit with the family and listen to their future plans on tree planting. In these visits we learn more of farmer-to-farmer exchange of ideas. From these meetings we plan for the meeting which takes place in September or October where new farmers register.

We have a demonstration plot at our premises to show farmers different practices and types of trees. We also have a central nursery to provide seedlings to our farmers. Because of high demand we are decentralizing nurseries, to schools for example.

Field days are always organized to show and teach the farmers of the advantages that can be delivered from engaging in some of the agroforestry practices. The picture and slide shows proved to be a good method at field days. It arouses farmers' interest in particular practices, especially when some of the slides are from the project area. A farmer recognizes a fellow farmer on a picture or a slide and knows that it can happen in his or her own area. The farmer sees this as a challenge and wants to take it up.

Networking

The project staff have a lot of contact with other organizations addressing similar practices. We have support from the government departments and from local NGOs. Project staff are usually invited to attend workshops and seminars within and outside the country on issues related to agroforestry. Through this project Family Farms is a member of other networking organizations which hope to learn from what others have achieved.

Project problems

Procurement of seed can be difficult for NGOs because the Forest Department usually supplies departmental nurseries first before they consider others.

Lack of co-ordination with other organizations and departments working in the same area implies need for a combined effort to assist the farmer.

Transport is a major problem in this area, and there is little co-operation between development organizations. Donor agencies are not co-operative about their staff assisting staff of another project. Limited resources are not used to the best advantage, for example some camp officers received bicycles from two different donors because there is no co-ordination between organizations.

Survival rates of trees are low, especially certain species such as *Leucaena sp.* and *Eucalyptus sp.* Such trees do not stand weed competition and moisture stress, and this makes them more vulnerable to attack by termites.

Very few women used to attend our meetings. Traditionally men attended. They left their wives at home so that they could be greeted on return, and in turn tell their story. This attitude has changed. Women have asked the project staff to conduct separate meetings. They also asked to have more meeting places to shorten the long distances covered.

Conclusion

The changes in extension methods which the project has made within the short period since its inception show the importance of flexibility to handle problems as they arise. Appropriate methods to address the problems of the farming community are explored.

The awareness, implementation, adoption, and monitoring stages have been achieved in our project area. Everyone associates Family Farms with trees. Many farmers register the implementation of some agroforestry practices through demonstration farmers, for example many farmers have started to leave *Acacia albida* in their fields.

It is important to note that the personal relationship between a farmer and extension agent is critical to the adoption of a technology. The way you present yourself to the farmers is of equal importance. You may try to approach bottom-up, but can end up presenting yourself in the old opposite style.

A Study of the Effects of Land Use on Water and Soil Resources on the Slopes of Mount Kenya

HANSPETER LINIGER

Introduction

The tertiary volcanic cone of Mt Kenya with a base diameter of 80 to 100km and the highest peak of 5200m asl is an ecologic island within a semi-arid environment. It provides water and fertile soils for agriculture and forestry for a great number of people. The river water resources have an important role in supplying perennial streams to dry areas up to several hundred kilometres away from the peaks of Mt Kenya. Conservation strategies by Kenyan authorities have so far protected those resources quite effectively. But recently, water resources are getting more and more limited (Decurtins *et al.*, 1988) and the soil erosion hazard seems to have increased (Liniger, 1991b). A growing demand owing to fast population growth and immigration, and fast changes of land use, increase the pressure on the resources (Kohler, 1987). This shows the need for maximizing the utilization of the water and soil resources in a sustainable way.

The immigrants come from high potential areas to a less favourable environment, where they lack the knowledge of making the most productive and sustainable use of the limited resources. They often have little or no means for any investments. Thus, they are restricted to small land areas and to options which require low inputs and little change from their cultural practices. Their situation is a daily struggle to satisfy their basic needs, and to avoid being caught in the poverty trap.

The Laikipia Research Programme (LRP) is a collaborative venture of the Universities of Bern and Nairobi. It is managed by the University of Bern and financed by the Swiss Development Co-operation.

Aim and general approach

Research objectives

These are to maximize the utilization of limited natural and human resources to satisfy the basic needs of the people, and to maintain or improve the natural resources. This involves the following steps:

- To assess the present resources and their utilization
- To identify human constraints in the present land use
- To identify the influence of human activities including land use on the resource conservation and utilization and productivity
- To identify areas of resource loss or mismanagement
- To identify low-input options for improved land use and resource management, and
- To assist the government and the local authorities in land use planning,

by predicting the consequences of different land use scenarios on the resources and the productivity.

Method used
The technique applied comprised:

- Three different area scales – the regional, the local, and the farm level. This different level approach shows the dependency between the neighbouring areas, for example the lowlands are getting resources from the neighbouring highlands
- Priority to low-input methods that require little cultural change, show short-term benefits and maintain or improve the natural resources
- Studies by postgraduate students
- Inter-disciplinary approach with emphasis on good field work, and familiarization of the researchers with the local population and their environment, and
- Long-term commitment of the donor, and local and foreign institutions.

A regional approach

One of the regional studies was an investigation of the water balance of the Naro Moru river basin from the peak of Mt Kenya at 3600m asl west to the semi-arid savannah at 1800m asl. The objective was to measure in different zones of the basin:

- The river flow during dry and wet seasons
- The recharge and discharge of ground water
- The amount of abstractions from the river, and
- To link changes in the hydrology to changes in land use.

Findings
The main ones were:

- The upper forest zone (3000 to 3400m) is the area contributing most surface runoff and groundwater recharge
- The lower forest zone, with deep volcanic soils of high water-holding capacity is a storage buffer and regulates the flow, and
- Below the forest zone abstractions deplete the flow.

Land use changes threaten to upset the pattern. Deforestation on the upper slopes is not yet severe, but encroachment into forest is observed. Increasingly, intensive small-scale cultivation is leading to larger and more frequent floods. The buffering action of the lower forest is dependent on permanent ground cover, and is threatened by increased cultivation.

Below the forest zone is an area of rapid change in land use as a result of migration from over-populated high rainfall areas. The subsequent change to densely populated small-scale farming results in increased demand and greater abstractions. At the same time, inappropriate land use appears to increase the water and soil loss, and reduce short-term and long-term productivity. A regional strategy is clearly required for the management of the water resources in these interdependent zones, with limitations and regulations on abstractions which take note of the needs of lower areas.

A site-level study

The use of water resources was studied at site level in an area with recent land-use change by immigrant farmers from high-rainfall areas, at two sites in the semi-arid highlands west and north-west of Mt Kenya. On crop land the local practice of interplanted maize and beans was compared with tied-ridging, mulching, and agroforestry systems.

Results from runoff plots showed that soil losses averaged 10 to 15t/ha/yr, and runoff up to thirty per cent of rainfall during the growing period. Tied-ridging was not effective. However, mulching, and agroforestry plus mulching, signficantly reduced the soil and water loss. Soil moisture is a critical factor in this area and maize yields increased from slightly more than 1t/ha under the local practice to more than 2t/ha on the mulched plots, mainly due to reduced evaporation loss.

The implication is that present practices are certainly causing reduction of potential production and long-term degradation, but the damage can be substantially reduced by low-input biological measures.

The extension message

Studies were made of farmer reaction to proposed new methods, partly through questionnaires, and also at workshops attended by farmers, agricultural officers, and project research staff. There were a number of emerging conclusions. First, farmers and agricultural officers strongly under-estimated the importance of water loss by evaporation in this semi-arid environment. Second, they recognized the value of mulching, but were concerned about the availability of material. Third, they were concerned about the competition in agroforestry systems (for example, alley cropping and live fences).

On all these topics, the perception of farmers and agricultural officers was similar to that of the project staff which was based on their trials. However, on the practice of minimum tillage there was a marked difference of opinion. The farmers and ministry staff felt that minimum tillage increased compaction, reduced infiltration, lowered fertility, and was laborious, whereas research showed exactly the opposite. The reasons for this were studied further, and it transpired that the immigrant farmers were judging the method on their background of high potential areas, where soil fertility improves with deep digging (getting nutrients from the subsoil into the top soil). It showed that farmers had little experience in the new environment. It also emerged that farmers showed some reservations towards 'doing a lazy job' by only scratching the surface for minimum tillage. They felt that this method was against their professional pride.

Some of the large-scale wheat farmers on the dry footslopes north of Mt Kenya have changed their cultivation practice over the past 10 years to minimum tillage and mulching using special light tined equipment. Their experience was very useful in the discussions with small-scale farmers.

This example shows the importance of communication between the farmers, extension staff, and the researchers. Only through this continuous exchange and interaction will research efforts become applied.

Conclusions

There is need for applied research to determine the amount of abstractions in the different zones, and to quantify the influence and consequences of land use change on the water and soil resources. Furthermore, a participatory approach is needed to involve the different communities along the catena. This is especially useful because of big distances, new immigrants, different cultures, and different administrative areas. Participatory approaches alone, without applied research, will not give the appropriate answers. This is because the people are either new immigrants coming from a completely different environment, or they are being settled on small plots, with no solutions to cope with this new situation.

The regional approach needs to identify the strategies for improved land and resource management for all the different zones. Otherwise the poorest at the end of the queue will not get anything. Two strategies seem to be very important. First, to improve the land use in the high-potential zone to maximize production and minimize the loss of productivity, and to assure the continuity of their function as the resource area for low-potential zones. Second, to make people in the low-potential areas more independent of the richer neighbours by improved resource management. Applied research has an important role to play in putting these theories into practise and providing the required information on the resources and their utilization, and the influence of the land-use, and also in bringing together communities, project implementers, administrators, decision-makers, politicians and researchers. The final goal should be to produce an overall strategy accepted by all the partners concerned. Because of the problems in defining sustainable development, applied research has to monitor the processes and to keep a flexible approach.

Several reports on this project are in preparation. Details can be obtained from LRP.

Developing Sustainable Grazing Associations in Lesotho

J.P. HUNTER and L.C. WEAVER

Introduction

Rapid population growth and slowly increasing off-farm employment have placed ever-increasing pressure on already over-used, communally-managed, natural resources in Lesotho. Rangelands are particularly hard pressed. This paper assesses the sustainability of one effort at developing community-based range management, the Lesotho Range Management Area programme (RMA).

By tradition, land is owned in common and managed by the King. In practice, this management is delegated to chiefs. Once powerful, their authority has steadily eroded over the years and their attempts at management are often ineffective. Government has adopted a variety of land husbandry and grazing control regulations in an attempt to supplement the chiefs' authority, but these also are rather ineffective. In consequence, Lesotho's once highly productive rangelands are now suffering from a rate of overgrazing estimated to be between thirty to fifty per cent. The result is soil erosion and low productivity.

Opportunities to improve rangelands and livestock are constrained by weak management and the free rider problem, both frequently encountered with communal resources. Relatively free access, combined with low range-use costs, stimulate livestock holders to expand their herds to a point where long-term sustainability of the range resource is threatened. According to rural surveys, farmers are not indifferent to low range or livestock productivity, but range tenure provides them with few incentives to do much about it.

The Range Management Area programme

Begun in 1982, with assistance from USAID, the RMA programme has sought to overcome these ecological and social constraints by setting aside an area (RMA) for the exclusive use of a number of range-using communities. The right of exclusion is seen as a prerequisite for residents to gain a sense of management control and ownership of the rangelands they use. The programme has set itself three long-term goals. These are, increased livestock productivity and income, greater commercialization of the livestock industry, and improved management of rangeland resources. These goals have been elaborated by several specific management objectives developed by RMA residents themselves. These include improved range quality, higher-valued livestock products, and more marketing opportunities for livestock and livestock products.

Management of RMA rangelands is undertaken by resident livestock owners through a grazing association (GA). A management committee made up of elected representatives and traditional authority provides overall direction to the GA. This committee selects from among its members a seven-member executive committee to handle day-to-day concerns. Overall technical guidance on range and livestock management and marketing matters is provided by a government-employed RMA adviser. Although he or she handles the day-to-day management of the RMA, the role is an adviser, not a manager.

Since 1982, four RMAs have been established, of which Sehlabathebe is the oldest. They encompass approximately six per cent of all rangelands in Lesotho, and have a population of about 18 000 persons. Livestock numbers include more than 16 000 large stock and 82 000 small stock. Plans are underway to begin two more RMAs within the next 18 months, and a total of six to eight more within the next 10 years. The programme hopes to have approximately twenty per cent of Lesotho's rangelands under RMA management at the end of this period.

A grazing management plan for the Sehlabathebe RMA was begun in 1983. By 1990, its rangelands had benefited from seven years of organized management. The grazing plan has sought to provide deferred rest in the mountain cattlepost areas, and continuous summer rest in the village grazing areas. Comparisons undertaken between 1983 and 1990 show improvements in total ground cover, range condition, range species diversity, and range trend. In addition to these improvements, there have also been improvements in livestock productivity, including a higher proportion of females in the cattle herd, improved cattle weight and condition, and increased livestock marketings.

These accomplishments have lent an aura of success to the RMA programme. USAID has pledged its support for a further 10 years and, significantly, other projects with range management components have decided to emulate the RMA model. However, despite this apparent success there are some long-term problems which, if not overcome, could seriously compromise the future of these efforts.

Problems of sustainability

Although some of these problems have been recognized, the relevance of others has not been sufficiently appreciated and few, if any, measures are being taken to counteract them. Still others are only potential problems at this point, but problems which may be realized as the RMA model matures. In what follows, reference is made to the Sehlabathebe experience unless otherwise noted.

Sustainability of the range management plan
Part of the initial improvement in range quality resulted from a thirty per cent reduction in stocking rates brought about by the exclusion of non-residents' livestock at the initiation of the RMA. This was a one-shot action, however, and no mechanisms have been implemented for capping livestock numbers. Officials sought to avoid a head-on confrontation over this issue. They hoped a project-instilled emphasis on livestock productivity

would replace desires for quantity with concerns for quality. However, pressure appears to be building up from two sources to increase stocking rates in response to improved range condition. On the one hand, former non-resident users, who had been debarred from the RMA, are seeking to have their animals readmitted. On the other, residents may seek to increase the size of their holdings.

To date, the number of ear-tagged cattle owned by residents shows no long-term upward trend. Data on small ruminant trends are not known, although the general feeling among project and government officials is that the trend is upward. If this is confirmed by subsequent data, it could compromise the effectiveness of the grazing plan.

Another potential problem for the viability of the range management plan is the possible contradiction between the goal of increasing the commercialization of livestock herds, and the desire to maintain and extend the already-achieved gains in range condition. In addition to wool and mohair production, which is already commercially motivated, officials have sought to promote increased marketed offtake. They hoped that this would increase incomes and relieve the stocking rate, increase culling, and improve livestock productivity. Some success in this area has been registered. If, however, increases in marketed offtake are to continue, one of four things must happen, alone or in combination:

- One, farmers must content themselves with smaller herds and be willing to exchange, one-and-for-all, animals for cash or some other asset
- Two, farmers must reduce other forms of offtake (theft, death, home slaughter) and channel the result to the market
- Three, the productivity of livestock herds must improve so that there are higher herd growth rates, and the subsistence functions of livestock can be undertaken with fewer animals, and
- Four, if the off-take rate remains unchanged, livestock herds must increase to permit higher marketings.

Evidence for these factors is at present ambiguous with no clear trends emerging. However, one, two and three are difficult to bring about, while four is likely to dominate. Thus, there is a danger that animal numbers will increase to accommodate the increased demand for marketed offtake, and further undermine the range management programme.

There also appear to be problems with the commitment of RMA residents to the rotational grazing programme. Although membership of a grazing association is not compulsory, all RMA livestock, whether owned by members or not, are subject to the grazing plan. The commitment of non-members to the plan may not be high. One might argue that this issue should not receive undue attention since the non-members have voluntarily opted out of the GA. However, there is a strong correlation between size of livestock holding (especially sheep) and GA membership, so things may not be as voluntary as they seem. According to several observers, although the vast majority of RMA livestock owners support the grazing plan in principle, a number do not support it in practice.

One investigator reported that non-cattlepost owners (who tend to have smaller holdings) tend not to practise the planned summer grazing rotation. Instead of rotating from lower to higher pastures as the summer progresses, they prefer to rotate in the opposite direction. Since many of these higher pastures formerly had been used by non-RMA residents, and residents' cattleposts were mostly in the lower pastures, one may speculate that non-cattlepost owners prefer the reverse rotation because it gives them first use of an ungrazed pasture, and a measure of freedom from largeholder domination.

Undermining adherence to the plan is also the fact that smallholders are often faced with a shortage of herding labour. As a result smallholders are especially concerned about stock theft and find it difficult and expensive to send their animals to summer pastures. Instead, many prefer to leave their animals in village grazing areas all year round. This is especially true of some small cattle owners who keep their stock primarily for the subsistence production of draught power and milk. This is not a problem now, but if it intensifies, village grazing areas may get insufficient rest during the summer to support planned stocking rates during the long harsh winter. As a result some large stockholders may be reluctant to bring their stock down to village areas during the winter.

If livestock holders are to voluntarily follow the dictates of the plan, they and their herdsmen must understand it. Despite substantial extension activity, questionnaires only demonstrate a rudimentary knowledge of the details of the plan, although understanding of its rationale is reasonably good. The separation of ownership and day-to-day livestock management may be the source of other sustainability problems. Most herdboys are adolescents with minimal education who receive minimal compensation. Their knowledge about, and commitment to, the task are often weak. This problem has concerned RMA personnel for some time, and steps have been taken to initiate training of herdboys. Results, so far, have been mixed.

Sustainability of the grazing association model
The Sehlabathebe GA has been from the beginning an organization of livestock holders, and special efforts were made to involve large holders in both membership and leadership positions. GAs in newer RMAs have continued this focus. Organizers justified a bias towards large holders on two grounds. First, in an attempt to ensure widespread acceptance of the grazing plan (large holders' opposition would almost certainly spell its doom). Second, because largeholders were thought to be more commercially-oriented and more responsive to extension advice. In the event, project officials were proved right. Largeholders have taken up leadership positions and the grazing plan has achieved wide acceptance and reasonably wide compliance.

The fact that smallholders are more likely to be in non-compliance, raises some questions about the long-term viability of this strategy. If the association appeared to be captured by largeholders, and turned into a club to serve their interests, its popular legitimacy, and that of the grazing plan, could be destroyed. Then the range could become the reserve of large stockholders and rather than reduce range use conflict, it could increase. Sehlabathebe provides some evidence for this. Despite a membership that has averaged about seventy-five per cent of all livestock owners, the members are disproportionately middle-aged males, formerly migrant but now resident, who rely on livestock for their primary or secondary income source. Non-members tend to be away on migrant labour contracts, subsistence producers, and female-headed households.

Counteracting this apparent bias towards largeholders are some membership benefits, mostly associated with the breeding programme, that are particularly helpful to smallholders. The GA's provision of bulls and rams gives smallholders access to good quality breeding stock. The association also provides grazing for cows being bred. This free benefit provides improved nutrition during pregnancy and lactation, improves reproductive performance, relieves village grazing areas, and reduces herding costs to the small producer.

Problems of long-term GA viability are intensified by the fact that non-holders of livestock, who make up approximately one-third of the households, are not targeted for GA membership. On the surface, this seems sensible. Range users should be responsible for regulating range use. However, the demarcation between livestock owners and non-owners, and between the livestock and non-livestock economies is not a clear one. In the first place, ownership is not a fixed state but a variable condition. Livestock accumulation is complementary to the prevailing migratory labour pattern. As savings accumulate from mine work, part is invested in livestock. Upon retirement, livestock provide income and can be gradually sold to meet cash needs. Thus, most non-owners are either young households at the beginning of the cycle or elderly households at the end. As potential owners, the first group should be able to represent their interests on the GA. As former owners, the latter group may provide valuable experience.

The division between the livestock and non-livestock economy is also blurred, particularly during the winter when livestock are grazed in village pastures. By custom, the stubble in harvested fields may be grazed unhindered. This makes it difficult, if not impossible, for farmers to plant fodder during the winter since they are unable to capture the benefits either for their own livestock or from sale to others. Thus fodder production, a key ingredient in any livestock productivity improvement programme, is retarded.

Enforcement of the grazing plan has posed further problems. Against non-RMA livestock this has been widespread and popular, although it has had to be continuous and there have been occasions of violence. Significantly, the GA's exclusion rights have been upheld by the courts. Against insiders, however, enforcement has been less vigorous. This is presumably because the GA wishes to avoid internal conflict. However, if this persists, its legitimacy and authority may be severely undermined.

Sustainability of the government input
The RMA programme does not occupy a clearly defined position on the top-down, bottom-up spectrum. In principle, it is intended to be at the

bottom-up end. Community-based range management is the purpose of the programme, and considerable effort has been expended in organizing farmers' meetings and associations. However, from the beginning, practice has not always lived up to intention.

Demand for the Sehlabathebe RMA did not come from the area's range users but from the Lesotho Government and USAID. Officials defined the programme goals, selected the technology package and management objectives, defined the RMA boundaries, and were instrumental in writing the GA's constitution and by-laws. Initially, popular participation seems to have consisted mostly of explanatory meetings by officials and chiefs. This has caused one investigator to term it a top-down project. This is not too surprising given that it was a pilot project to try out new techniques for managing communal resources. It would be a source of concern if there is no evolution from this approach. Efforts have been made to increase participation, but there is more to do.

The newer RMAs have avoided some of the shortcomings of Sehlabathebe. Request for RMA formation has come more genuinely from the residents themselves, and they have taken a greater role in delimiting RMA boundaries. As a result, boundaries have been drawn more on cultural and social grounds and less on a technological and ecological basis. Thus, some of the conflict with outsiders experienced at Sehlabathebe has been avoided. Development of the newer Pelaneng RMA has involved much greater participation by residents in the organization of the RMA, founding of the GA and its constitution and by-laws, and implementation of the grazing plan and technical package, than was the case elsewhere.

A related issue is the leadership dynamics of the association itself. The project has recognized the importance of providing leadership training for executive committee members. To this end, courses on simple book-keeping, record-keeping, and group decision-making have been conducted by Lesotho Distance Teaching Centre. This has undoubtedly improved the efficiency of the GA. It has not, nor was it designed to, overcome conflicts over personality and village or national politics within the group. This role has fallen largely to the RMA adviser. Because he is an outsider and is seen by the GA as technically competent he can function as an honest broker, adjudicating disputes and settling internal conflicts. Unfortunately, he has been provided inadequate training for this role. RMA advisers have training in range management and livestock matters, as they should have, but they have no formal training in the group dynamics and motivational skills that could help to strengthen the participatory model that government and project officials strive to support. Nor have they training in the management of routine, day-to-day project activities.

Sustainability of the livestock productivity technical package
The RMA programme's success hinges ultimately on its profitability to range users. This means that the cost and effort expended in forming and running the GA, the grazing plan, and all the ancillary activities of the RMA must provide tangible, worthwhile benefits in terms of higher livestock productivity and human well-being. The programme has not

addressed this directly. The recommended livestock productivity package of rotational grazing, improved breeding, culling, and animal health inputs may be the best one, but as yet no attempts have been made to test this. Neither the profitability to the farmer, nor a comparison with other packages have been assessed.

As a range management project, it was assumed that improved range condition would pay off in the long-run in terms of better livestock condition, improved fleece weights, and wool and mohair quality, and improved herd dynamics. No one really knew, nor did they collect the data necessary to find out. Since tangible short-term benefits are necessary to retain residents' support, several strategies that bore no obvious relation to the grazing plan were promoted, such as breeding to imported males, culling and animal health care.

How would the recommended package compare with one which tied improved herd dynamics, greater wool and mohair production, and improved draught productivity, to range improvement and fodder production for supplemental feeding of females and offspring? How cost effective is the importation of breeding males versus selective breeding from highly genetically-variable RMA herds? How cost effective are increased drenching and dosing? No one knew at the programme's onset, and no one knows now.

Conclusions

The development of grazing associations is an ongoing process in which continued development has been, and must be, based on lessons learned from earlier experience. The following conclusions emerge from experience.

From the beginning the programme has placed too much emphasis on the ability of an inadequately evaluated technical package to deliver sufficient benefits to justify the effort expended by RMA residents, government, and the donor. Reflecting the technical bias of government and project officials, it has over-emphasized the expected pay-off from range improvement, and under-emphasized its role as an input into a package of activities designed to improve livestock productivity and enhance human well-being. The current package should be properly evaluated, and compared to other options. RMA residents must play a leading role in this, and participatory appraisal techniques provide the best means for this. The result may be not only a re-appraisal of goals, but also a re-appraisal of the components of the technical package. Next, an economic appraisal of selected alternatives should be undertaken to determine their comparative economic viability.

Despite national commitment to a participatory model, this has been only partially achieved, and then unevenly. In its favour, the programme has made an attempt to understand the social dynamics operating within the RMAs. However, greater attention must be paid to expanding the mandate of the GA, and with it the membership base to include all residents, not just those who have livestock. This should include greater attention being paid to the special needs of smallholders and non-livestock owners. Finally, officials, particularly the RMA adviser, must be selected for their commitment to the participatory model, and receive training in group dynamics and motivation and participatory development.

The potential problem of increasing livestock numbers must be faced head-on before it becomes real and threatens the viability of the grazing plan. The proposed national grazing fee will be too weak to effectively constrain livestock numbers in the RMAs. Yet the RMAs should have both the will and the organization to implement a control programme on their own. This is not the place to discuss the details of such a programme, but it would be preferable if it were as self-policing as possible. Thus market, or market-like, solutions should be emphasized over prohibitions and fines.

If these issues are addressed, many of the weaknesses in the present model should be overcome, and the RMA programme should have a better chance of long-term success.

Conclusions

Next Steps Towards Better Land Husbandry

R.J. CHEATLE

Introduction

'Environment and the Poor' assembled a mix of senior development personnel, primarily from east, central and southern Africa, for the purpose of taking stock and reaching a consensus about the approaches required to secure conservation farming. Realization of the limited impact of development activities in stemming degradation processes, and improving the smallholder condition throughout the region, has given impetus to the search for better ways to apply investment resources.

Representing a range of disciplines and agencies in the public and private sectors, the experienced specialists participating at the workshop showed the effectiveness of inter-disciplinary collaboration confirming that important changes are under way, and providing a regional consensus about the development directions to take. This volume presents lessons from a wide range of participatory experience. It shows results from a workshop forum, which demonstrates convergent thinking in support of a participatory approach to improve land husbandry.

The workshop message

The workshop message can be summarized by two related concepts, one about the integration of conservation practices into farming systems, and a recommendation for a participatory approach to conservation farming. Derived from linking the two concepts is the idea that conservation farming can be achieved by professional agents who are 'working with farmers for better land husbandry'. A tentative definition of land husbandry, prepared by an Environment and the Poor working group, endorses this notion. It calls for a commitment to a participatory approach to conservation farming, to enable sustainable management of natural resources by rural people.

The definition is: 'Land husbandry is the implementation and management of preferred systems of land use in such ways that there will be no loss of the land's stability, productivity or usefulness for the chosen purpose(s). When appropriately managed, land can be used to satisfy human needs in the present without damaging its potential or restricting future options for alternative uses.

'In the context of small-scale farming, land husbandry involves land users making decisions to manage their resources of soil, climate, plants, animals, implements, labour, knowledge and capital. Improvements to land use and management practices should develop from the knowledge, problem awareness, analysis, ideas, insights, goals, aspirations and priorities of the farmers and their households. Decisions about changes to existing land use practices should involve the fullest participation of the land users in identifying problems and opportunities, formulating and implementing appropriate courses of action, and in monitoring and evaluating the results of doing so'.

This definition effectively draws conservation into the mainstream of smallholder development, by emphasizing improved land use and linking the concept of land husbandry with people's participation. What the definition does not do is create a link between conservation and markets, or between conservation and institutions. Crops that provide good ground cover or fix nitrogen in the soil will not be grown unless there are other benefits. This may require marketing arrangements or markets, and institutional collaboration.

Next steps for better land husbandry requires a co-ordinated activity in five fields:

- Improved land use
- People's participation
- Institutional development
- Policy and incentives, and
- Markets and marketing

Improved land use

Survival strategies of the rural poor do not include brave notions about land husbandry. Their concern is to scratch a living, fill bellies, get wood for fuel and shelter from where they can. To survive, people will mine common resources, and their own lands if necessary. Work with farmers is frustrating if their rights to land use are insecure. Given security of tenure to the parcel of land or rights to its use, the poor farmer will invest time, money and labour in it.

A partnership between farmer and professional can be useful, for knowledgeable professionals can provide good ideas, and facilitate input of resources to increase the rewards. In participation for better land husbandry there are critical processes set about the offering of an opportunity by an agent and its acceptance by a client. The agent to client process relationships are not well understood. But these are critical to development, because they are about sharing knowledge, and reaching agreements with clients about what to do to improve what is already being done. This approach contrasts with the unsatisfactory situation where the professionals decide what rural people should do.

In the 10 years to 1990 more than US$20 000 000 has been disbursed annually in Ethiopia as Food for Work. The money has largely been used to build terraces, bunds, and other physical measures on farmers fields in Ethiopia. Other donors in Ethiopia, and throughout the region, have committed funds to the old approach to reducing land degradation, the control of soil and water movement by physical structures. That approach has treated soil conservation as an activity separate from smallholder production.

In Ethiopia any success appears to be limited, and some measures may have done more harm than good. Engineered measures for soil conservation are unwanted by many people, and once put in at considerable cost they may be abandoned, neglected or removed, because people do not perceive advantage. When they cannot produce enough food for the family they do not want fifteen per cent of the small parcel of land taken up by bunds. Nor do they want bench terraces cut into shale that break away in the rains and are too narrow for turning the plough. Like Ethiopians they would probably build a terrace, take the food reward, and that would be the end of the matter, unless they were shown how to use the terrace to produce more food or make some other kind of profit. But, that has not been done. People respond to opportunity. It is the job of the development community to help identify sustainable opportunities, not quick fixes of a temporary nature. Today, in Ethiopia, and throughout the region, there is renewed interest to find ways in which investment resources can be applied more efficiently to improve land use by sustainable land husbandry practices. The workshop showed the potential for improving resource investment through a move from soil conservation to a whole-farm or land unit approach to conservation farming, where the emphasis is upon implementing good land husbandry practices to increase production or other benefits for the farmer.

A radical shift is required, towards prevention of degradation, by giving more attention to agronomic practices, and less to conservation engineering. Promoting organic matter recycling, good ground cover, techniques for getting more rainfall into soil, balanced land use, and good spatial relationships between land uses are important for maintaining fertility and preventing erosion. When a proper land use balance is achieved much of the need for structural protection of soil is eliminated.

That approach is also good for crops. For example, organic recycling improves conditions for plant nutrition, soil structure, and the moisture-holding capacity of the soil, while improved cover through the rainy season protects the soil against erosive rainfall. Assume a farmer monocrops sorghum. If crop residue can be returned and beans can be intercropped with the sorghum, the consequence is good conservation and better crops, so this integration of conservation with production produces benefit for the farmer at low cost. Good land husbandry combines sustainability with profitability.

Showing how to use conservation to produce benefits is critical to adoption of technology, and hence to stemming degradation. Adoptability is often overlooked, and this is another necessary condition.

The range of agronomic practices for conservation farming are well-known, as is their relative efficiency in conservation. Much can be achieved by rearrangements involving the existing plants and the manner in which they are grown. Adaptive research and demonstration designed to fit the local environment may be all that is necessary for a considerable advance. Planting fodder grasses under fruit trees is clearly an improvement to production and conservation. So is the planting of *Acacia albida*, and the initiation of contour ploughing on a field to grow sorghum instead of cultivations up and down the slope. The positive effects of the tree on the crop and more moisture in the soil will bring yield increases. The tree also increases biodiversity and product diversity. Clearly the benefit of additional sources of firewood and fodder, and a better crop, will reduce the farmers risk. We know what can be done but are less certain about how to go about fitting practices into a farming system to achieve adoptability.

People's participation

Experience recorded in this volume suggests routes to adoptability. First, outsiders must work with farmers and understand what they do, and why they do it. Understanding the farming system is a lead in to making positive adjustments to the production process, land use, and marketing. The social and economic facets of the system must be understood, as well as the physical. Promotion of a new cash crop may fail if the harvest coincides with a staple, or if the farmer can earn more money by selling his labour or weaving cloth on the farm.

Participatory rural appraisal (PRA) is reported as one of the efficient approaches whereby outsiders can obtain an understanding, rural people can make known their problems and aspirations to a multi-disciplinary team, potential solutions can be identified in partnership, and agreements can be reached about what to do. Appraisal also offers an excellent vehicle for obtaining a popular perspective about the results and consequences of what is done. Learning lessons is important.

The use of a multi-disciplinary team is advantageous, because people with different knowledge and skills can help people look critically at their own systems. Farmers and specialists can make their technical opinion about improvements to what is being done, and have that opinion subjected to critical consideration. This is a practical and quick way of determining what changes would be feasible.

After PRA comes the doing, or participatory technology development (PTD). Most often the work will be adaptive research or demonstration, and will probably require the investment of resources to determine an adoptable solution to the problem identified by PRA. The arguments for a PRA/PTD approach are compelling. For a whole-farm approach, good diagnosis, collaboration and linkage, the stimulation of local interest and the involvement of people in their development process are common threads in the local successes reported at the workshop.

Some examples

The work of Fr Gappa is a record of environmental rehabilitation over a long period of time. Acting as a catalyst he has stimulated local people to improve the resource base and their own security in terms of food, fuel, and shelter. The work of the Kenya Institute of Organic Farming has brought about advances in sustainable land use and production benefits in much shorter periods, notably with womens' groups. In the Catholic diocese of Kisii, also in Kenya, catalytic efforts by one or two persons is also showing evidence of positive impact in the short term. Working with farmers can produce good results, but all the successes have required considerable investment of resources in time and labour, though some of the NGO work has been done on modest budgets.

There are several points to draw out. First, good people willing to serve can help others to achieve without the need for major financial inputs, but these implementers have to be identified and supported. Much of the vanguard work for participatory development has been done through NGOs, and well-run organizations who wish to assist in relevant investigations offer opportunity for investment funding.

All too often, the lessons derived from project experience remain unpublished. This volume reports some of the good work but there is more to unearth. There is urgent need for sound ideas, and proof of practices that work well for the farmer, and for conservation. Some of these practices are customary. We know that they exist and offer potential for extrapolation, monitoring and evaluation. It is necessary to make use of the opportunities presented through investment, by evaluating the relevant project work to learn the lessons, and making the information widely available. The call is for improvements in monitoring and evaluation, for databases, and for information exchange by networking throughout the development community. But information is not just for the professionals. There are always land users who cannot be reached. They too should be made aware and offered any opportunity that others have adopted. More use of farmer-to-farmer exchanges will enable some to see the empirical evidence and to talk with farmer colleagues. When a farmer sees pumpkins on a neighbour's land twice as big as his or her own, the farmer will ask about how this is done, and learn to follow the example. A further opportunity for farmer-to-farmer extension lies with the media. Publishing a farmers' perspective, with advice and tips for conservation farming in a well distributed newsletter is likely to influence other farmers.

Collaboration between agencies, programmes and projects could lead into situations where better use is made of information, and others learn about the utilitarian things that farmers do. Whatever is done it is always too little, for manpower and finance are scarce resources. Our development approach should encourage more local responsibility, control, and management of the development process. Farmers do experiment and serve as extensionists. There is a need to work out ways by which more people can be assisted to help themselves. One thing is certain, there will never be enough professionals to service all the farmers.

Community development is an entry point to user-led development for better land husbandry. Having established a working relationship with a community, a multi-disciplinary team can initiate participatory rural appraisal (PRA) to enable sensible local decision-making about development issues, to assure a development plan and to inculcate local ownership of that plan. But PRA is more than this. It is an orderly approach to problem definition, to the design of development activities to solve problems, and to the securing of agreements about what to do. Through investigation of the totality of a community situation the participatory effort can effectively design interventions to fit the local situation. Once this is done it should be possible to devolve much of the responsibility for agreed work to local people.

Responsibility is most likely to be with an interest group, perhaps a womens' group or a church related mixed group. Group work decided by people for themselves is likely to continue until a result is obtained, whether this is adaptive research, demonstration or some other aspect of development. Exciting possibilities are offered. Improved efficiency in the use of manpower will be derived if a number of established groups were assisted by a community agent, with technical backstopping by a team of specialists. Second, there is the possibility of applying financial resources to established groups, thereby reaching the goal of placing resources at the grassroots, perhaps through a village development fund.

Another plus for in-depth appraisal is the prospect of improved project conceptualization and design. All too often projects are initiated without rigorous preparation of objectives and a pre-implementation evaluation.

Institutional development in the public sector

A participatory approach for conservation farming can be made to work, but so far there is little evidence of it being applied in the public sector, where much of the resources and manpower lie. There is an opportunity to mobilize there, and perhaps a responsibility to encourage investigations of the new directions. There is interest and debate in the public sectors about participatory development, but multi-disciplinary teams have not been fielded, and inter-disciplinary collaboration is still a conceptual toy. Development of a cross-sectoral planning approach to conservation farming is hindered because public agencies are structured in compartments and geared for top-down operations. The soil conservation department helps build the bunds, while the livestock department deals with the fodder that grows on them. An approach based on inter-disciplinary collaboration is not easy for them.

The implications of a participatory approach are simply enormous. Major changes are required in the functions, structures and procedures of institutions – no easy task to accomplish. Fielding and funding an inter-disciplinary team for a one-off operation is difficult, as a contribution from Zambia shows. Institutionalization is a major challenge. These problems will have to be overcome slowly, through long-term projects, and the donor community can help through precise orientation of funding over the longer term. There is an urgent need for action on the ground to develop the role models that will provide real measure of the potential for the participatory approach to conservation farming.

Where there is interest to investigate the potential of participatory development there is the potential for action and funding through a project frame. Role model projects will attempt to do two things, through a series of steps. As there is no model of participatory development for conservation farming that can be applied widely by the public sector, the methods and approach at the technical level have to be identified and proved. This work requires the fielding of an appropriate multi-disciplinary team, its funding, and a mandate for the team to let them get on with the job. As the business of participatory development requires various disciplines to work with people at the grassroots, the members of the team will be derived from several departments or agencies, so it will also be interdisciplinary. This means that the group cannot be accommodated readily within a department or even a ministry. That team will have to take new directions for which there are no organizational precedents. It seems likely that the answer may have to be an operation funded and managed as a pilot operation outside the confines of institutional structures.

The future looks towards a technical approach based upon conservation farming in which second echelon teams with an area responsibility practise PRA and provide technical backstopping to PTD. For the ongoing development activities a community agent may well service a number of self-help groups. Implementing such arrangements at the field level does require the creation of a generalist, the community agent, and an improved focus upon groups rather than individuals. The functions change too, for work arises because of problem definition through a dynamic system of appraisal at the local level.

As well as a technical model there is also need for an institutional model. This is a chicken and egg situation for a model cannot be achieved without changing the institutions. But a role-model project can pave the way for institutional change by conducting an investigation about institutionalization of the technical model. Practical examples can be produced to demonstrate how best the bits of the model can be placed in an institutional setting. Therefore, the role-model project has two jobs – method development at the technical level and an investigation of needs for institutional change and development. In practice, both jobs could proceed in parallel. The establishment of a high-level land husbandry working group to operate in concert with the field level activities would provide support to them, and an opportunity for senior people to reach a consensus about needed changes and to work for them. The role would also be to present the information that would influence policymakers, and the development of a favourable policy climate for participatory development is a fundamental step in its progress. Major policy changes without good evidence of efficacy in the proposal are unlikely.

Policy and incentives

The state has a responsibility to create the conditions which encourage land users to practise productive forms of sustainable land use. The situations are always complex and there are no single-step solutions. However, governments have the authority and responsibility to develop strategic plans, identify priorities, direct resources to the poorest of the poor, and to support development of a methodology for a participatory approach to conservation farming. Participatory development initiatives are required for conservation farming, and high-level support for them is needed.

On other fronts, there is need for policy support for land husbandry and all facets of conservation land use. Insecurity of tenure is often an underlying cause of degradation. Farmgate prices of agricultural crops are often pitifully low, and the identification of a plan to combine conservation with production to produce benefits can be difficult. We have learned that regulation for soil conservation does not work well. What is needed now is a package of policy incentives to encourage conservation farming. One that encourages people to invest in their land resources and enables independent rural people to sell their products in a free market.

Policy changes do not come without acceptance of argument for change, and lobbying. The need for that argument is part of the rationale for role-model projects about conservation farming. A forum to present argument is also useful. That is why the need for a director-level workshop to discuss policy issues was identified at Environment and the Poor.

Markets and marketing

While the first concern of the rural poor is food security, most farmers are acutely aware of prospects in the local market-place. Perceiving a market niche where there are good profits to be made the farmer will try to respond, but may be unable to act because he or she cannot get the product to market. Smoothing the way to market and profit have not been subjects at the fore in soil and water conservation activities. With conservation largely dependent upon what the farmer does in the fields, the conservationist must take an interest in the market-place. If the conservation opportunity lies with sweet potatoes because of their ability to provide good ground cover quickly, the conservationist needs to know if there is a market for the crop, and perhaps what can be done about improving it. Conservation opportunities must be set in simple but direct economic terms, so that farmers can see financial benefits. Here there is a useful role for an economist, to provide simple budgets for conservation options that contrast and compare the relative advantages of one option versus another.

In this volume are examples of crops with conservation attributes for which there is market demand in Europe and Africa, and the potential of a high net margin for farmers. Some of those crops also offer opportunities for value-adding activities through processing. For these crops there would seem the strong possibility of combining profitability with sustainability and adoptability. It is to be hoped that some pragmatic conservationists will try to promote such prospects and work with farmers for better land husbandry to their profit.

Improved Training Approaches for Sustainable Land Husbandry

K.H.M. SEGERROS and R.J. CHEATLE

New training for a new approach

As a participatory approach for conservation farming develops there will be a need to make substantial modification of training approaches and curricula to create new attitudes, skills and awareness within professional people and farmers. One effect of moving from top-down to bottom-up will be to create retraining needs at all levels.

Much good work has been done. We have a basket of technical practices that are known to be effective in controlling soil erosion and water. In most cases we know what to do, and the next step is how to put it into operation in farmers' fields.

Recognition of problems has stimulated a debate and initiated a transition in roles, attitudes and approaches. In the technical areas, the shift is towards prevention of degradation by seeking ways that farmers can change their agronomic practices to combine production with conservation. Resource management professionals and decision-makers are beginning to understand that their role in the new agenda for sustainable resource management is changing from regulator and ruler to facilitator. Their job is to help farmers identify and tackle their own resource management problems.

The manpower shortage

Africa's trained scientific manpower is about one quarter the relative strength of Asia. This situation is not improved by increases in school roles and declining per capita spending on education. Without the necessary resources to disburse on improvements, schools produce relatively few outstanding performers to enter tertiary education. A graduate manpower gap exists in the region. There is urgent need for highly-qualified specialists with an orientation in sustainable land husbandry, such as rural anthropologists, economists with training at the micro-level, as well as the established range of technical disciplines. While this cadre of well educated professionals will help accelerate the changes underway, their work to develop fresh training approaches and methods will require programmes fitted to all levels of activity.

The needs of formal training

Formal training within institutions is the main approach to public sector staff training. In these institutions, text books, teaching materials and curricula at agricultural training institutions are rarely up to date, and seldom address the needs of small-scale farmers. At a typical institution there are soil conservation students. They are taught such things as Manning's formula, design criteria for V-flumes, and how to interpret aerial photographs in formal classroom lessons. At the institution they may gain experience in the practical construction of broad-based terraces on two to eight per cent slopes using theodolites and heavy machinery. It is rare for students to gain practical experience with farmers on the steep slopes that many farmers use, in conditions where tractors, theodolites and aerial photographs are unavailable. An approach to education and training is needed that teaches students how to think for themselves, improvise and bring them into regular contact with farmers. Implied is a need for field-based training.

The right curriculum

In terms of the curricula, there is urgent need for changes so that the student can learn about conservation farming and the new approaches in land and water management on the ground. On the job the ex-student rarely encounters those ideal conditions of two to eight per cent slopes. Farmers may be cultivating seventy per cent slopes in semi-arid conditions. Though this practice may be forbidden by conservation regulations, population pressure or arbitrary land allocations puts the farmer in a position where he or she has no alternative. This is the reality, and students need appropriate preparation on farms, where they can make do with line levels and hand-labour, like the farmer.

Practical training on-farm, consistent with the needs of small-scale farmers will provide the orientation and practical experience needed. In this way students will learn the basics of their trade from the farmers and studies of on-farm situations. Without hands-on field experience no one can learn the skills of participatory rural appraisal. Specialization cannot be entirely neglected, and neither can the classroom. There needs to be a balanced programme of training that provides each student with opportunities to study areas of interest in depth. All cannot be generalists, and everyone should have a store of relevant knowledge. Today, the primary need is to scrutinize and review curricula and training approaches. Extension specialists, researchers, private entrepreneurs, farmers' representatives, as well as academics and educators, should be involved in this exercise, to identify, and tailor, forms of education and training that prepare students to meet the demands of the market.

Bridging gaps in collaboration

From the initiation of training the need is to provide for interaction and collaboration between student and farmer, and also between student and potential colleague. At work, the agronomist on the other side of the office corridor is likely to have an extensive knowledge about hybrid maize monocropping on large farm fields, but very limited knowledge of smallholder intercropping with open pollinated maize, and virtually no knowledge about either conservation or integration of conservation practices into the production systems.

Typically, contact between subject matter specialists will be minimal, yet, as in the example, the agronomist and conservationist need to relate their skills and knowledge. Unfortunately, neither one nor the other has been prepared for collaboration in the field, or the realities found there. Unprepared, they cannot contribute well to the development of sustainable land husbandry. All too often, both stay in the office and write reports. Good training will take into account the need for collaboration and knowledge of immediate relevance by bringing people into these practical roles from the start of their experience. Training in issues, such as the development of a collaborative work programme and communications, is just as important as teaching for acquisition of technical knowledge.

Ways of providing the necessary field experience include the setting up of bases in the field, from which students under tutelage can work with farmers on development issues. Secondment to ongoing projects is another.

Training targets

The future land managers
If school children are trained and informed about proper land management techniques the prospect for increases in sustainable production will look brighter in the future. School teachers require training in sustainable land husbandry. Part of their job will also be to bring land husbandry into the school curriculum through creation of specialist subjects and the strengthening of existing subjects with material about conservation. Training material and teaching approaches will have to be developed to meet the requirements of these changes.

Farmers
At the village level, farmers' training should go beyond technical issues. It is important that farmers should actively participate in the formulation of the training, to ensure that their priorities and needs are adequately addressed. Communities should be encouraged to implement their own training needs assessment as part of participatory rural appraisal and community-led appraisal. Encouragement of this kind of initiative will require flexibility on the part of the professional community, as well as a commitment to react positively to requests. Farmers could do much more in applied research and the reporting of it, in training each other, in learning through farmer-to-farmer exchanges and networking.

Extension agents
Putting the farmer first is not easy. We are still locked into old attitudes and ways of working. The farmer's first and most frequent point of contact with the extension system is with the extension agents. These frontline staff often have limited formal training. But they do have one particular advantage, they can interact with farmers on a daily basis. This potential

capacity for acquiring local knowledge and determining the needs of farmers is very often neglected and rarely acknowledged. Yet, extension agents should be critical facilitators in development activity, and efforts should be made to stimulate and motivate them. This can be done in several ways, for example:

- Train extension agents in how to listen and learn from farmers
- Encourage improved reporting of field conditions and responses. A guiding principle is that local information should be gathered, collated, analysed and acted upon
- Provide relevant literature, research support, communication and training material. All too often it is impossible to find a book or technical magazine at an extension office
- Organize regular in-service training in participatory development and sustainable land husbandry
- Develop a career structure with conditions of service to attract good staff
- Develop PRA for routine investigations and monitoring, and thereby involve researchers and senior extensionists in regular collaborative work with frontline staff and farmers
- Involve researchers, senior extensionists, frontline staff and farmers in the production and analysis of project reports, and
- Involve researchers, senior extensionists and frontline staff in collaborative studies of information flow and technology adoption. These are vital but neglected subjects.

Subject matter specialists
All too often the potential of subject specialists is under-utilized. Supervision of field staff, organization of field demonstrations, development of area specific recommendations and other duties are neglected. Part of the problem can be alleviated by well designed in-service training programmes. Areas with need for development and implementation are:

- On-farm technical courses
- Courses in communication techniques
- PRA courses, to encourage collaboration and participatory development
- Introductory training in socio-economic subjects for technical officers
- Introductory training in technical subjects for socio-economists, and
- Attachment of new officers to an experienced colleague, so as to learn on the job.

Researchers
Researchers need to link with small-scale farmers and identify the critical problems faced by farmers. Traditional soil and water conservation practices need to be improved, and opportunities for alternatives and additional complementary practices identified. At present there is too much concern with runoff plots and other basic research issues that do not help farmers or extensionists at all.

Researchers are a valuable human resource, and should be involved in education and training, notably in respect to very specific technical inputs to meet identified needs at the field and second echelon levels. Through participation in regular staff meetings, community-level meetings, and studies by extension personnel or students, researchers can stimulate informal training through contributing to these interactions and these roles need strengthening. Examples of subjects for collaborative study are:

- Successful resource management systems
- Unsuccessful resource management systems
- Traditional techniques and approaches to soil and water management
- Perceptions of land degradation at different levels of society
- Sensitivity to degradation within different agro-ecological zones, and
- Risk avoidance, uncertainty and innovative behaviour.

Policymakers
Education for policymakers also needs to be addressed. Land husbandry is a policy issue and resources need to be applied to meet training and other needs. Farm management includes social, cultural, economic and political aspects, as well as the techniques of production. The major need is for awareness creation, through pithy reports, seminars, well-designed field visits and other kinds of training experience designed to fit each local situation. Policymakers need to be brought into the field for well-designed keynote discussions. Creating opportunities for awareness creation will brief, and therefore enable, decision-makers to provide the strong support needed for work to promote sustainable land husbandry. Fortunately, it is numerous decision-makers, at this time, who are in the vanguard of participatory development.

Meeting training needs

It will be very difficult for each country to develop and implement all the training programmes needed to meet future demands. A good deal of training will have to be on an informal basis. Every job description for NGO and public sector staffers should include a requirement for contributions to training. Whenever there is work in the field there is a training opportunity, whether this is on a one-to-one or group basis, or whether it is for five minutes or a day. People can and should learn from each other as they work. Farmers can teach, and professionals can teach. The requirement is to encourage the habit.

A strong formal training sector will continue to be needed, but many countries will find difficulty in providing adequate resource allocations, even when arguments for need are precisely stated. In-country one of the most urgent needs at this time is for participatory in-depth reviews of training, complemented by an assessment of training needs, followed by analysis and development of priorities, strategies and tactics.

Regional training programmes
In order to support the new research and development agenda the primary human resource need is for a cadre of professionals to serve as trainers

and developers of methods. Their roles are to train others, and to help in the development of training programmes, methods and forms of organization appropriate to participatory development. As no single country can tackle all the problems, the implementation of creative training strategies and tactics on a regional basis would be cost-effective, utilitarian, and enable the development of quality programmes. Several core institutions could be involved. This is not to imply a need for new buildings necessarily but for an appropriate form of shared and neutral management. In the early stages these institutions would emphasize the training of trainers and the development of training methods.

Federated management, or institutional sharing has not been successful historically. Arguments over the location of programmes, financing, ownership of assets, and allocation of management positions, and national jealousies led to disbandments. Revival of shared facilities will require a fresh approach, one where the users meet on neutral ground as equal partners in an enterprise. Attempts are underway to develop regional programmes based within specific countries. These programmes and the participating trainees from third party countries will have to become operational under precise terms.

A benefit of regional training is regional co-operation. This will arise through the development and management of specialized national institutions, the resolution of needs for staff exchanges, co-ordination and harmonization of entry requirements, curricula and qualifications, and representation of faculty boards. There will be need for:

- Prioritization, and development of strategies and tactics to meet real needs
- Development of common curricula for participatory development
- Development of on-farm training methods
- Setting of common entry standards
- Equivalence of certificates, diplomas and degrees
- Staff development programmes
- Promotion of seminars and courses on specific topics
- Organization of staff, student and farmer exchange programmes, and
- Establishment of a scholarship award scheme.

Towards new approaches in training

The first initiatives are being taken and it is expected that more will follow. Kenya and the SADCC block provide two examples. Within Kenya there are several small projects with a training component in participatory development, and others researching the technical aspects of sustainable land husbandry in the public and NGO sectors. The SADCC Environment and Land Management Sector has commenced work to review training needs and develop a training programme. This work will bring together knowledge and experience gained in the past, and then is expected to move into a phase of integrated programme formulation, with an emphasis on practical field training and a sustainable land husbandry orientation. The programme will include short studies, training courses, and training for village people in model areas. Through the work in villages particularly, and the other components, it is hoped to stimulate new ideas, demonstrate their validity and generate useful feedback to channel into new initiatives.

Appendix

The Workshop Background

The workshop entitled Environment and the Poor was conceived as a regional activity in Addis Abeba at the 1989 meeting of the International Soil Conservation Organization. Discussions there initiated interest and support for the notion that an inter-disciplinary group of professionals should be brought together to provide a perspective for a way forward in development. A perspective designed to impact at the grassroots through integration of conservation with smallholder production. In Kenya an inter-disciplinary group of professionals began to organize the workshop. This example was soon followed by a group in Tanzania, and later the SADCC group of countries was represented by their Environment and Land Management Sector. The Permanent Presidential Commission for Soil Conservation and Afforestation in Kenya, and the Ministry of Agriculture, Livestock Development and Co-operatives in Tanzania agreed to act as the senior national hosts. Both local organizing committees were supported technically by the World Association of Soil and Water Conservation, SADCC, and the World Resources Institute. Numerous donors provided the necessary funding. The distribution of a first circular produced a wave of interest from almost every country in north-east, east, central and southern Africa.

The interest was there because the timing was right. The first circular asked for subjects to be discussed at the workshop. The response demonstrated a widespread recognition of the need to review methods and approaches, and to take a bold step forward in environmental management. Professional people in the region had become aware of the need to explore the social and economic issues in order to learn how to identify and encourage the adoption of sound land husbandry practices for smallholder production systems. At the same time, there was concern to link conservation practice with popular participation and rural appraisal, and discuss the utility of these concepts to sustainable development. The purpose of the workshop was to bring together senior professionals to exchange ideas and experiences about issues they regarded as important, and to encourage them to work as a think tank to produce a consensus. Inter-disciplinary participation was organized in order to ensure a common voice for the range of disciplines, for public sector and NGO development agencies, and for research and extension.

Both organization and implementation of the workshop worked well and helped prove that inter-disciplinary activities can work and be effective. The organizing committee had members from the relevant public sector implementing agencies, NGOs and universities, and this pattern was repeated in the workshop participation. In the working groups people worked together, and learned a lot from each other. The most important features of the workshop are the convergence of thinking that enabled senior and experienced people to reach a powerful regional consensus. This consensus provides a sense of direction for development. Summarized, it suggests five things.

First, we must investigate the potential for improving smallholder production systems through an integration of better land husbandry practices into production systems to combine conservation with good business practice.

Second, we must develop and refine methods and practice for achieving good land husbandry by participatory development, so as to give the land user a voice and create a working relationship. This would facilitate improved practices, which can be fitted to meet real on-farm needs, and take advantage of opportunities there and in the market-place.

Third, we must use participatory appraisal for project design, and develop the tools and methods to monitor and evaluate development activities.

Fourth, we must take steps to provide the training to create awareness of trends, needs, and to fit people for the new roles.

Fifth, there must be a process to review and modify the functions and structures of institutions to enable effective inter-agency collaboration and operations in a participatory mode. To provide improved guidance, and prove participatory approaches, key role-model projects should be developed, brought to public notice, and studied in depth. In this way the approaches and methods can be refined, lessons learned, and the useful information disseminated. An outline of the next steps proposed is set within the conclusion of this book.

Reporting the workshop

The source material for the book is the papers presented and the reports of many discussion sessions. The exchange of ideas, and the realization that a major step forward in rural development is now possible, generated such a powerful surge of enthusiasm that the organizers decided it had to be turned into a book, produced at low cost to be affordable to all those interested in rural development.

The workshop had three components. First a small group of sociologists and anthropologists met for a few days in Nairobi, to exchange ideas and consider how to present them at the main meeting. For the second part, 92 participants from 22 countries met at Arusha in Tanzania from 2–5 June 1991 to gain practical field experience of participatory rural appraisal. The third part was held in the Aberdare Mountains, close to Nyeri, Kenya from 6–11 June, and included the presentation of papers and working group discussions.

So much material was presented at the workshop that the editors had to select and shorten the papers included in this volume. The editors thank the authors for understanding and tolerating the sometimes brutal editing necessary to keep within the target size and cost of the book. Some papers refer to other published material about the project being reported. Readers seeking more information are invited to write directly to the authors at the addresses given in the list of participants and authors. There is also a list of further reading.

Acknowledgements

The workshop received magnificent support from many agencies in the public and private sector, notably in Kenya and Tanzania. The local organizing committees worked hard to make this workshop a success. The help of the committee members and those who gave them official support to work on this project is gratefully acknowledged. Agencies approved staff time and accessed physical resources for this project. The Chief Executive of the Permanent Presidential Commission for Soil Conservation and Afforestation in Kenya, the Principal Secretary of the Ministry of Agriculture, Livestock Development and Co-operatives in Tanzania, and the Director of the SADCC Conservation Unit, have been active and supportive throughout.

Many individuals contributed to the Environment and the Poor workshop, and are recorded in the list of participants and authors. Thanks are due to authors and to the participants. Without the written contributions and the critical appraisal in working groups nothing would have been achieved.

Special thanks are due to certain people. From the initial conception Dr Lill Lundgren, then working with SIDA, gave her personal support and arranged the early funding that got things started. Professor Norman Hudson and Dr William Moldenhauer acted as advisers, put in a great deal of time and provided valuable advice. Post-workshop Mr H. Kimaru, Mr M. Mbegera, Mr P. Mung'ala and Mr S. Lyimo continue to be in the vanguard of those active in exploring the workshop consensus.

Without the donors listed overleaf there would not have been a workshop, or a book. The large number of donors who came forward with money and technical advice is indicative of the considerable interest in the workshop subjects. They are thanked on behalf of all those affected by their support.

THE DONORS

The Commonwealth Secretariat
The Ford Foundation
The German Agency for Technical Co-operation (GTZ)
The International Development Research Centre (IDRC)
The Japanese International Co-operation Agency (JICA)
The Norwegian Office of Regional Aid and Development (NORAD)
The Overseas Development Administration (ODA)
The Rockefeller Foundation
The Southern-Africa Development Co-ordination Conference (SADCC)
The Swedish International Development Authority (SIDA)
The Swiss Development Co-operation
The World Association of Soil and Water Conservation (WASWC)
The World Resources Institute (WRI)

Opening Address in Tanzania

DR B. MOSHI
Principal Secretary, Ministry of Agriculture, Livestock Development and Co-operatives, Tanzania.

Soil and water management problems are fundamental to many smallholder production systems. They present a formidable challenge, one that must be addressed to ensure food security and enable satisfactory levels of production, all within a framework of a sustainable environment. Agricultural systems in many parts of the third world are characterized as complex, diverse, risk-prone and are under tremendous population and other pressures to increase present production at the expense of long-term sustainability. A more realistic sustainable agriculture would be more in tune with the local resource base, make maximum use of internal production inputs, and have potential for sustained production and profits further into the future.

While sustainable agriculture may be a societal goal, the practices that will provide sustainability can only be implemented at the farm level, where they must form a sustainable farming system. In the mid 1980s there was an increasing recognition of the central role of farmers in implementing sustainable agriculture. This has produced a changing view of the farmer's role in agricultural research, from client to partner. In Tanzania the farming systems research and extension (FSR/E) approach was developed in order to address the agricultural development needs of resource-poor farmers. Among the basic tenets were building on the good points of the existing farming system, and taking into account the farmers' indigenous technical knowledge. The FSR/E approach has tried to address the problem in several ways and areas, and we are pleased to note some success in their endeavour. Today, there is renewed interest in bottom-up approaches utilizing participatory rural appraisal and multi-disciplinary and inter-disciplinary team approaches to resolve on-farm problems.

Sustainable agriculture relies on renewable resources that do not contaminate the environment, and that can be afforded by resource-poor farmers. Sustainable production systems which integrate conservation can be made to pay off for the benefit of the smallholder. Yet there is much to be done to create awareness, to provide appropriate forms of organization, and to identify and assist adoption of further technologies. Experiences in many countries have shown that one of the major reasons for low adoption of developed agricultural technologies by farmers is the inappropriateness of the technologies to their physical, biological, and socio-economic conditions. In order to identify and refine soil and water management technology, so that adoptable choices emerge, there is need for popular participation in the research process to ensure that research activity has a sharp focus. To ensure relevancy at the farm level, it seems logical that the implementors of a sustainable farming system, that is the farmers, must be involved in the research leading to appropriate technologies.

Opening Address in Kenya

MR J.T. ARAP LETING
Permanent Secretary and Head of Public Service, Kenya

In most developing countries, there has been renewed interest in mobilizing the community in environmental conservation matters. In Kenya, we are fortunate to have a leader who is committed in environmental conservation matters. His Excellency the President has personally led us in constructing soil conservation structures in various places in the country. Barely two years after ascending to the presidency, he established a Permanent Presidential Commission on Soil Conservation and Afforestation (PPCSCA) to co-ordinate national programmes in soil conservation and reafforestation. Many environmental conservation programmes have benefited from the work of this commission during the 11 years of its existence.

The 1972 Stockholm conference on human environment and habitat marked the turning point by focusing the world's attention on the environmental degradation crisis. This led to the setting up of the United Nations Environment Programme (UNEP) based in Kenya, and renewed commitment in many other international agencies in combating environmental degradation. It is encouraging to note that even product-oriented research centres such as the International Potato Centre (IPC), International Livestock Centre for Africa (ILCA), International Maize and Wheat Improvement Centre (CYMMT) have joined hands with the International Union for Conservation of Nature, the UNEP, Food and Agricultural Organization (FAO), the World Resources Institute (WRI), and the International Council for Research in Agroforestry (ICRAF), in promoting sustainable environmental conservation.

Despite all the efforts to improve the performance of soil and water conservation programmes, there are numerous examples of programmes which have failed, in spite of using the latest available technology. The main reasons for this poor performance include:

- Farmers were considered to be part of the problem rather than essential parts of the solution. Consequently, they were completely left out of the decision-making procress to the detriment of most conservation programmes
- Environmental matters have been relegated to lower levels in the list of urgent national concerns as the debt burden facing developing economies increases
- Lack of commitment on the part of the local farmers who cannot perceive any benefit from environmental conservation effort, and consequently have no interest or motivation to conserve
- Lack of sufficient political and financial support to environmental conservation programmes. In some cases the necessary rules and regulations are in place, but there are no adequate mechanisms to enforce them
- Inadequate co-ordination of various agencies involved in environmental conservation matters, both within government and outside
- Lack of adequately trained and experienced people to generate and disseminate information on environmental conservation matters, and
- Lack of infrastructural support and appropriate land tenure systems.

The sponsors and workshop organizers have brought together people with different experiences, and the multi-disciplinary discussions will lead to practical solutions to the constraints identified above. The investment made through deliberations into environmental protection is investment in a more secure future for ourselves and generations to come. The success of such an investment will be judged by the appropriateness of the recommendations, and the follow-up actions, initiated to turn this region into a greener, safer, and more productive environment.

Closing Speech in Kenya

MR C.R.J. NYAGA
Director of Forestry, Ministry of Environment and Natural Resources

In order to protect our already existing sensitive and fragile ecosystems, particularly the natural tropical forests and other biospheres, increased use of inorganic fertilizers is essential to sustain soil fertility level. This, together with soil and water management, will ensure acceptable food security without compromising social political stability.

These issues must be addressed within the context of the global interdependence, based on real partnerships by developed and developing countries in development of man and environmental protection – in other words the alleviation of poverty, environmental protection and conservation.

In developing strategies and methodologies to assist the poor it must be remembered that the majority of rural people, poor or marginally well-off, are women, children and old folks. There are a number of immediate implications:

- Participatory integration of rural people in decision-making
- Community mobilization through collective effort, co-ordination and integration of diverse ideas, needs and objectives
- Retraining of contact persons, for example, extensionists and planners
- Changes and reorientation in institutions of decision-making towards rural poor's participation in decision-making and framing of strategies, and
- Strategies or decisions to bring research messages to the poor to incorporate practical wisdom of rural farmers into research designs and processes.

It is my sincere hope that recommendations will not end up on shelves in offices and libraries, but will be interpreted into practical applications at farm level. I urge the governments and donors to help the implementation of these worthwhile recommendations. Implementation of them will go a long way in improving the production and sustainability of the smallholder farming systems, in the conservation of our soil and water resources, and to the alleviation of poverty as one of the most noble of human ideals.

Let me take this opportunity to thank the donors for their continued commitment to the conservation of our natural resources and for funding this workshop. Contribution from the Kenya and Tanzania governments was highly appreciated. Mr Chairman, I would also like to thank WASWC for implementing some of the sixth ISCO meeting recommendations – and for accepting Kenya and Tanzania's invitation to host and organize this workshop. I am sure that WASWC will continue to assist us in developing strategies for combating environmental degradation and in the implementation of the recommendations.

The effort and time put into planning, organizing and execution of such a workshop is enormous. It requires dedication, patience, tolerance, and co-operation to bring together such a unique gathering. This would not have been possible without such an able team of organizers both in Tanzania and Kenya. Such a success deserves a very special word of thanks and encouragement. Keep it up!

Last, but not least, I would like to thank all the participants for their active participation that led to the development of such practical recommendations.

The Role and Contribution of Rural Sociologists to Sustainable Soil and Water Management
Statement by the Workshop Participants

Context

The trend is rapid and accelerating change in approaches to soil and water management, with increasing emphasis on farmers' participation. Farmers' resistance to past soil and water conservation programmes is now more understood to be the result of the design and implementation of programmes without farmers' participation, and without sufficient analysis of their social and economic constraints and opportunities. Recent rapid methodological developments with participation present a big challenge to sociologists and anthropologists. The purpose of this statement note is to explore, and outline, the potential roles and contributions of sociologists and anthropologists in participatory soil and water management.

Contribution – constraints

Sociologists and anthropologists are an important resource for soil and water management programmes because they are trained to deal with social and cultural aspects of development. They have approaches and methods which facilitate interaction with, and learning from, rural people and communities. They are able to probe beyond bare facts to reveal underlying causes, and understand and appreciate peoples' sentiments, feelings and attitudes. Some soil and water management related factors which are often overlooked or misinterpreted are land tenure and tree tenure, labour availability, management and use of common property resources, livelihood strategies, household decision-making, gender issues, local perceptions of costs and benefits, and peoples' time-horizons (for example short-term and long-term views on planning).

For applied work in soil and water management, sociologists and anthropologists leaving universities need practical experience in field applications. Also, applied sociologists are still scarce, and often marginal in their organizations. In consequence, the potential contribution of sociologists and anthropologists to soil and water management programmes has as yet been largely unrealized.

Implications

For employing agencies and donors:

- Sociologists and anthropologists should increasingly be involved in agencies, projects and programmes dealing with soil and water management, both as short-term consultants and on a long-term basis
- Where possible, sociologists should be employed as sociologists in their professional roles
- There should be a close link between the permanent sociologist or anthropologist and the short-term consultant, to ensure continuity and follow-up of the sociological input
- In order for sociologists and anthropologists to have an impact and to be able to provide professional support and guidance to upcoming sociologists in agencies and programmes, it is vital that senior posts be available to them, with a clear definition of roles and responsibilities
- Sociologists and anthropologists should be available and able to contribute effectively at all stages of the project cycle, from problem identification and planning, through implementation and monitoring, to evaluation
- Senior sociologists and anthropologists employed in agencies and programmes should participate in in-service training of professionals and practitioners in other fields. The objective of such training is to encourage the use of simple applications of sociological and anthropological approaches and methods. It is also to make other professionals understand when, and for what purposes, a professional sociological input is required
- The potential role of the anthropologist and sociologist as a cultural broker between so-called experts and beneficiaries of projects, that is as intermediaries and interpreters of the respective world-views and interest, should be made use of, and
- Sufficient time and funds should be allocated for the sociologist and anthropologist to be able to do a professional job.

For universities and colleges:

- Students in sociology and anthropology should have field attachment and experience as part of their training
- Training institutes should advocate the importance of applied sociology for practical development work, thereby helping to increase the demand and to clarify the roles and functions of the applied sociologist or anthropologist in the project and agency setting, and
- Training institutes should make efforts to keep contact with practising sociologists and anthropologists, give them professional back-up and

support, and organize in-service training. This will ensure a professional standard and identity can be maintained. A closer link between sociological training and research institutes, and practising sociologists and anthropologists, will also help to improve training and stimulate research which can be applied in practical development work, including soil and water management. At present, such research is often discouraged as unprofessional in the academic world.

For sociologists and anthropologists:

- Sociologists and anthropologists should make themselves visible, taking the initiative rather than waiting to be called upon. This will help increase the demand for their professional input
- Sociologists and anthropologists should establish networks (regional or national, or both), to keep contact, share experiences and learn from each other, and
- Sociologists and anthropologists should keep up with, further develop, and spread participatory approaches.

The Role of Economics in the Sustainability of Smallholder Systems

J.P. HUNTER and N. REYNOLDS

Role of the economist

Economists in land management are concerned with the way rural people procure their livelihood from the land, and the incentives and constraints that promote or retard sustainable land use. Since society's resources are necessarily limited, economists are also concerned to evaluate resource utilization in an effort to promote those uses with the highest social payoff. While technical viability is certainly a necessary condition for a land management practice or intervention, it is not sufficient. The practice or intervention must also be economically viable, that is, it must make recognizable economic sense to land users.

Economists work with a basic premise about human behaviour. It is that in most activities involving choice, but especially those which affect their livelihoods, humans rationally choose those options which advance their interests most. Although economists recognize that custom, tradition, and a variety of other motivations may affect behaviour, when they conflict with economic rationality, they require special explanation. This simple premise has proven to be analytically powerful and predictively useful.

From the perspective of land management, the premise implies that existing practices, even though they may be unsustainable over time, are rational in the context of the existing incentives and constraints faced by land users. Farmers and herders do what they do, not because of ignorance, superstition, or wilfulness, but because it makes sense in terms of their priorities. Therefore, land management interventions must change the incentives and constraints so that land use becomes compatible with sustainability.

Thus, a key task of the economist, is to study the farming system, both to understand its internal rationality and to identify ways in which constraints can be relaxed and the incentive structure altered to promote good land husbandry. Revised arrangements for credit, subsidies, taxes, and tenure can affect land use directly. Other fiscal measures related to old-age income security, wage labour opportunities, exchange rates, and the operation of parastatals have indirect land use implications. On the basis of an in-depth understanding of land users' rationale for economic activity and allied sociological assessments, economists, social scientists, technical specialists and land users can work together to design more sustainable and efficient practices.

Economists, or other professionals with interest and experience in economics, should be involved from the very beginning in research, project or activity design, and implementation, to ensure that the activities proposed are economically-viable and capable of being adopted by land managers. This involves evaluation of the technology, analysis of farm budgets, consideration of input and output markets and credit facilities. Economists should also be involved in on-going monitoring and evaluation of projects. In a participatory mode, economists can assist with the identification of farmers' priorities, and in the calculation, both objectively and from the community's perspective, of the valuation of costs and benefits. This function is important since the valuation given by farmers can be very different from that provided by economists unfamiliar with their priorities. Sustainability must be especially concerned with farmers' own valuations.

Integrating economics into development activities

The use of economists in soil and water management is at present quite limited for several reasons. First, there is a general shortage of economists with training and experience in resource management economics. This is especially the case at the field level, and also often applies to the headquarters level. Second, technical specialists in resource management positions often do not understand how economics might be used to improve their effectiveness. They often see economists as trying to block or delay technical initiatives without fully understanding economists' reasons for doing so. Third, economists have not always comprehensively explained their functions in land management to their less economically-tutored colleagues. This problem is compounded by the fact that economists tend to have a vocabulary unique to themselves and arcane methodologies. Often, they present their findings in forms (equations, graphs) not easily understood by laypeople.

Initiation of a new dialogue between technicians and economists is now vital if economists are to be appreciated and their skills fully utilized.

Economists are frequently seen as being too distant from activities on the ground to understand local priorities or problems, and their valuations of costs and benefits are often perceived as arbitrary, abstract and unrealistic. Resentment exists about the perceived power of economists in the project planning and implementation cycle – they have the final word, they control funds, they are arrogant and first among equals. Economists accustomed to top-down action need rehabilitation to work effectively in a participatory environment. Technicians also need education and training in economics, so that they can introduce economic tools and analysis into their work.

Greater receptivity to economists' input into activities can only occur when there is a more widespread understanding and use of economic principles in the normal course of development work. Economists must devise ways of bringing the essence of their analysis and thinking to non-economists in simple, straightforward ways. Since economics deals with the day-to-day activities of people, this should not be too difficult. However, it does require some special thought. In addition, economists must familiarize non-economists with economic ways of solving economic problems. Experience in Zimbabwe and elsewhere suggests that, if training is carefully done, villagers and members of development teams can become economically-literate with little effort. Once all parties work mutually with economic concepts, they should begin to exhibit greater confidence in their ability to dictate their desires to outside experts. Economics can then become a means by which local concerns and priorities and outside expertise are integrated. For this to happen economists must adopt a more co-operative mode of behaviour, not only with their colleagues on the planning team but also in their relations with resource users. They must recognize that local communities and professional colleagues can often help them to identify priorities, assign values to resources, and to assess project costs and benefits.

Conclusion

In addition to changes in perspective and practice towards a participatory approach, efforts must be made to increase the number of resource management economists and land management practitioners with a background in resource economics. This will require a long-term commitment of universities, agricultural colleges, and resource management departments, since resources will have to be allocated to sending people away for further studies and to adding or upgrading the resource economics component of existing curricula.

An encouraging aspect of the development process is the increasing recognition that economists and economics are necessary components of sound land management practice. Now, economists must find ways to work with land users and professional colleagues more effectively, especially in regard to investigations of the potential of participatory development.

Participatory Approaches to Soil and Water Conservation for Sustainable Smallholder Development
Statement by the Workshop Participants

Until recently, most smallholder soil and water conservation projects and programmes have been top-down, that is programmes and technologies conceived and developed centrally, and then transferred to the field. This has often taken the form of inflexible programmes, with a heavy emphasis on earthworks, and little choice of practices offered to farmers. The results have frequently been disappointing, with lack of farmers' enthusiasm for implementation, and poor maintenance of physical structures. Participation, where it has occurred, has often been a case of the professionals gather data, analyse it, prepare plans, and then ask farmers if they agree before requesting mobilization of local resources, including labour, to implement these plans. Farmers and other resource users have had limited opportunity to be actively involved in the development and decision-making processes of project design, and even less in policy formulation. The recent rhetoric has been participatory, but the reality has often remained top-down.

The new participatory approaches to soil and water management, including participatory rural appraisal (PRA), other methods and tools, are bottom-up and have specific applications in soil and water conservation activities. Group discussions and reports have indicated a growing convergence of new thinking and a general agreement on the importance of participation. There is an increasing convergence as to the potential of PRA and similar methodologies to improve soil and water management, with implications for changes in policy, research practice, training, and regional collaboration.

These methods and tools have been evolved and tested in a number of countries, including Kenya where the term PRA was first used. PRA has been used primarily by professionals to work with farmers to collect and analyse information on ecological, socio-economic, and political circumstances for project design. There have been some attempts to use the information collected through PRA to strengthen the negotiating position of communities with their governments and the development assistance community. Information obtained by PRA has also contributed to national and sub-national environmental planning and policy formulation.

Some of the main elements of these new ideas and approaches entail shifts of emphasis **from**:

- Promoting soil and water conservation **to** improving land husbandry
- Ensuring soil and water conservation through increased regulation and restriction on farmer behaviour **to** an increasing emphasis on lifting local constraints to enable farmers to achieve sustainable soil and water management

- Conserving soil and water by physical structures **to** maintaining and enhancing soil and water management and productivity by improved agronomic practices
- A single-sector approach to project design **to** a multi-sectoral and multi-disciplinary effort
- Starting with the knowledge and technologies for soil and water management of professionals **to** starting with the knowledge and existing technologies of farmers
- Professionals lecturing, holding the stick, and presenting their ideas **to** listening and learning, handing over the stick, and encouraging farmers to express their ideas
- Data collection and analysis, and planning primarily by professionals **to** incorporating data presentation, analysis and planning by farmers, with professionals as facilitators (community consultants)
- Extracting information from farmers using standardized questionnaire surveys, **to** an array of participatory methods of learning from, with, and by farmers. This includes the use of participatory mapping, village landscape modelling, transects, analytical diagrams, matrix ranking and scoring, and action planning
- Identifying priority needs and options by professionals **to** the identification and selection of priorities by farmers and other local resource users, with assistance from outside technical expertise
- Blanket recommendations **to** a so-called cafeteria of demonstrated choices offered to farmers for them to test, evaluate, and select, and
- Technology development on-station to vertical transfer down to the farmers, **to** a more balanced research programme involving the legitimization of farmer experimentation, improvement of existing and indigenous soil and water management practices, and technology development by and with farmers on their fields, with the intention of lateral transfer between farmers.

Experience to date has demonstrated that participatory approaches and methods can be popular, powerful, and cost-effective for work with farmers to identify priority needs. They lead to identification of appropriate and sustainable options for the farmers to implement. It also seems likely that they may result in more sustainable and self-reliant soil and water management at the local level. They are not a panacea, but early results are sufficiently promising for the methodologies to deserve further spread, testing, adaptation, and development in different countries and contexts. For land management this has implications with regard to:

- Continued participatory methodological research and development (including participatory policy formulation, participatory monitoring and evaluation)
- Further research and evaluation comparing these new participatory approaches and methods with the conventional
- Dissemination of information to extension staff, researchers, decision-makers in government and the development assistance community

- Exchanges of information, methods, and experiences in these methodologies
- Development of training material and programmes, the sharing of trainers between countries, regions, and continents, and
- Institutionalization of participatory approaches in the public and private sector agencies.

Ensuring participation in the developmental work for a sustainable environment will require a fundamental shift in the professional approach. The initiation of activity is now expected to be with farmers to identify and act, on local priority need, opportunities, and options. Already, this move toward increased local participation is reflected in the changing philosophies and operational practice within development agencies. These trends are to be encouraged.

References

Abel, N.O.J., Drinkwater, M.J., Ingram, J., Okafor, J. and Prinsley, R.T., *Guidelines for Training in Rapid Rural Appraisal for Agroforestry Research and Extension, Amelioration of Soils by Trees Programme*, Commonwealth Science Council, London 1989.

Ashby, J.A., Quiros, C.A. and Rivers, Y.M., 'Farmer Participation in Technology Development: Work with Crop Varieties', in *Farmer First, Farmer Innovation and Agricultural Research*, Chambers, R., Pacey, A. and Thrupp, L.A., (eds.) Intermediate Technology Publications, London, 1989.

Brokensha, D., Warren, D., and Werner, O.L., *Indigenous Knowledge Systems and Development*, University Press of America, Lanham, Maryland, 1980.

Bunch, R., *Two Ears of Corn: a guide to people-centered agricultural improvement*, World Neighbours, Oklahoma City, 1985.

Chadhokar, P.A., 'Area closure for soil conservation', in *Proceedings of the 6th ISCO Conference*, Addis Abeba, Ethiopia, 1989.

Chamala, S., and Mortiss, P.D., *Working Together for Land Care: group management skills and strategies*, Australia Academic Press, Brisbane, 1990.

Chambers, R., Pacey, A. and Thrupp L.A., (eds.), *Farmer First, Farmer Innovation and Agricultural Research*, Intermediate Technology Publications, London, 1989.

Cheatle, R.J., 'Tree Growth On A Compacted Oxisol', *Soil and Tillage Research*, 19, 1991, pp 331–344.

CIMMYT, *Planning Technologies Appropriate to Farmers, Concepts and Procedures*, CIMMYT, Mexico City, 1980.

CIMMYT, *Towards the 21st Century: CIMMYT's Strategy*, Mexico DF, Mexico, 1989.

Conroy, C., and Litvinoff, M., *The Greening of Aid: sustainable livelihoods in practice*, Earthscan Publications, London, 1988.

Conway, G.R., 'Agro-ecosystem Analysis', *Agricultural Administration* Vol. 20, 1985, pp. 31–55.

Conway, G.R., McCracken, J.A. and Pretty, J.N., *Training Notes for Agro-ecosystem Analysis and Rapid Rural Appraisal*, second edition, International Institute for Environment and Development, London, 1987.

Decurtins, S., Leibundgut Ch., and Wetzel, J., *Resources Of River Water in Eastern Laikipia – River Water Potential And Rural Water Supply In Eastern Laikipia (Kenya) For Small-scale Farming*, Laikipia Reports No. 12, Bern, 1988.

Douglas, M.G., 'Integrating Conservation into Farming Systems: the Malawi experience' in *Conservation Farming on Steep Lands*, W.C. Moldenhauer and N.W. Hudson (eds.) SWCS & WASWC Ankeny, Iowa, 1988.

Douglas, M.G., *Integrating Conservation into the Farming System, Land Use Planning for Smallholder Farmers, Concepts and Procedures*, Commonwealth Secretariat, London, 1989.

Douglas, M.G., 'Integrating Conservation into the Farming System, Land Use Planning for Smallholder Farmers', in *Development of Conservation Farming Systems*, Report of the Training Course in Agoo, La Union, the Philippines 25 June–21 July 1990. ASOCON Report No. 4, ASOCON, Jakarta, Indonesia, 1991a.

Douglas, M.G. *The Development of Conservation Farming Systems: some policy and institutional considerations*, paper presented at the Workshop on Conservation Policies for Sustainable Hillslope Farming, Solo, Indonesia, 11–15 March, 1991b.

Douglas, M.G., and Lai, K.C., *Conservation Project Design. Report of the Training Course Held in Masvingo Zimbabawe 1–26 August 1988*, Report 19, Soil and Water Conservation and Land Utilization Programme, SADCC Co-ordination Unit, Maseru, Lesotho, 1988.
EHRS, *Highland Reclamation Study Vol. 1*, Final Report, FAO project UTF/ ETH/ 03/ETH, FAO Rome, 1986.
FAO, Organic Materials As Fertilizers, *Soils Bulletin 27*, FAO, Rome, 1975.
FAO, 'Soil and Water Conservation for Semi-arid Areas', *Soils Bulletin 57*, FAO, Rome, 1987.
FAO, *The Conservation and Rehabilitation of African Lands – An International Scheme*, ARC/90/4, FAO, Rome, 1990a.
FAO, *An International Action Programme on Water and Sustainable Agricultural Production*, M/U1108/E9.90/1/2500, FAO, Rome, 1990b.
Farrington, J., and Martin, A., *Farmer Participation in Agricultural Research: a review of concepts and practices*, AAU Occasional Paper 9, Overseas Development Institute, London, 1988.
Francis, C.A., and Hildebrand, P.E., 'Farming Systems Research and Extension (FSR/E) in Support of Sustainable Agriculture', in *Contributions of FSR/E Towards Sustainable Agricultural Systems*, proceedings of the 1988 Farming Systems Research/Extension Symposium, University of Arkansas, 1988.
GRZ, *The National Research Action Plan*, Research Branch, Department of Agriculture, Lusaka, 1990.
Grandstaff, Somluckrat, W. *et al. Rural Systems Analysis*, report on an international training workshop held in north-east Thailand, April–May 1990, Southeast Asian Universities Agro-ecosystem Network, Khon Kaen University, Thailand, 1990.
Gypmantasiri, *et al.*, and Conway, G.R., *An Inter-disciplinary Perspective of Cropping Systems in the Chiang Mai Valley: key questions for research*, Multiple Cropping Project, Faculty of Agriculture, University of Chiang Mai, Thailand, 1980.
Harrison, P., *The Greening of Africa*, Earthscan Publications, London, 1987.
Hjort af Ornäs, A., and Salih, M., (eds.), *Ecology and Politics: environmental stress and security in Africa*, Scandinavian Institute of African Studies, Uppsala, 1989.
Hudson, N.W., *Soil Conservation*, Batsford, London, 1981.
Hudson, N.W., 'Soil Conservation Strategies for the Future', in *Land Conservation for Future Generations*, edited by Sanarn Rimwanich. Proceedings of the 5th International Soil Conservation Conference. Department of Land Development, Bangkok, Thailand, 1988.
Hudson, N.W., *A Study of the Reasons for Success or Failure of Soil Conservation Projects*. Soils Bulletin 64, FAO, Rome, 1991.
Hudson, N.W., *Land Husbandry*, Batsford, London, 1992.
Hussein, J., *Agro-climatological Analysis of the Growing Season in Natural Regions III, IV and V of Zimbabwe*. Proceedings of a workshop entitled Cropping in Semi-arid Areas of Zimbabwe. Vol. 1, August 1987, Agritex, Harare, 1987.
ICRAF, *Annual Report 1990* ICRAF, Nairobi, 1991.
ILEIA, *Participatory Technology Development in Sustainable Agriculture*, ILEIA, Leusden, The Netherlands, 1989.
International Federation of Agricultural Producers (IFAP), *Sustainable Farming and the Role of Farmers' Organizations*, CTA, Wageningen, The Netherlands, 1990.
International Fund for Agricultural Development (IFAD), *Soil and Water Conservation in Sub-Saharan Africa: issues and options*, Rome, Italy, 1986.

Jones, E., Nyamudeza, P., and Busangavanye, T., 'Rainfed Cropping And Water Conservation And Concentration On Vertisols In The South-east Low-veld Of Zimbabwe', in *Vertisol Management in Africa*, IBSRAM, Bangkok, 1989.

Juma, C., *Biological Diversity and Innovation: conserving and utilizing genetic resources in Kenya*, ACTS, Nairobi, 1989.

Khon Kaen University, *Proceedings of the 1985 Conference on Rapid Rural Appraisal, Rural Systems Research and Farming Systems Research Projects*, University of Khon Kaen, Thailand, 1987.

KIOF, *Field notes on organic farming methods*, KIOF, Nairobi, 1990.

Kohler, T., 'Land Use In Transition, Aspects and Problems Of Small-scale Farming In A New Environment: the example of Laikipia District, Kenya', *Geographica Bernensia*, Africa Studies Vol. A5, Bern, 1987.

Kokwe, M., 'The Role of Dambos in Agricultural Development in Zambia', in *Wetlands in Drylands: the agro-ecology of savanna systems in Africa – Part 3e*, ed. Scoones, I., IIED, London, 1991.

Liniger, H.P., 'Water And Soil Conservation For Short-term Survival In The Semi-arid Footzone North-west Of Mt Kenya – Consequences For Soil Productivity', *Mountain Research and Development* Vol. 8, No. 2 and 3, 1988.

Liniger, H.P., *Water And Soil Conservation And Utilization West To North Of Mt Kenya – Concept And Some Results Of Applied Research For Sustainable Resource Development*, Paper presented for the international workshop on African Mountains And Highlands – 'Dynamics Of Resource Systems And Its Consequences For Sustainable Development', Rabat 19–27 September 1990, Morocco. Proceedings to be printed in *Mountain Research And Development*, Colorado, USA, 1990.

Liniger, H.P., 'Agro-ecology And Water Conservation For Rain-fed Farming In The Semi-arid Footzone West And North-west of Mt Kenya' in *Ecology And Socio-economy of Mt Kenya Area*, Winiger, M. et al. (eds.), proceedings of the workshop on 'Ecology And Socio-economy Of Mt Kenya Area', March 1989, Nanyuki, Kenya, 1991a.

Liniger, H.P., 'Water And Soil Conservation In The Semi-arid Footzone North-west of Mt Kenya', in *Erosion, Conservation, and Small-scale Farming*, Tato, K., and Hurni, H., eds., University of Bern, Switzerland, 1991b.

MacCormack, C., and Strathern, M., (eds.), *Nature, Culture and Gender*, Cambridge University Press, Cambridge, 1980.

Mascarenhas, J., et al., *Participatory Rural Appraisal*, proceedings of the February 1991 Bangalore PRA trainers workshop, RRA Notes, No. 13, IIED, London, 1991.

McCay, B.J., and Acheson, J.M., (eds.), *The Question of the Commons: the culture and ecology of communal resources*, University of Arizona Press, Tuscon, 1987.

McCracken, J.A., *Participatory Rapid Rural Appraisal in Gujarat: a trial model for the Aga Khan Rural Support Programme (India)*, IIED, London, 1988.

McCracken, J.A., *Training Programmes for Rural Women in Sustainable Use of Natural Resources*, Commonwealth Secretariat, London, 1990.

McCracken, J.A., Pretty, J., and Conway, G., *An Introduction to Rapid Rural Appraisal for Agricultural Development*, IIED, London, 1988.

NES, *Participatory Rural Appraisal Handbook*, National Environment Secretariat, Kenya, Egerton University, Kenya, Clark University, USA, and World Resources Institute, USA, 1990.

Nicou, R., and Charreau, C., 'Mechanical Impedance Related To Land Preparation As A Constraint To Food Production In The Tropics', in *Priorities For Alleviating Soil-Related Constraints To Good Production In The Tropics*, IRRI, Manila, 1988.

Pagiola, S., *Preliminary Estimates On The Economics Of Soil Conservation In Kenya*, PARD Soil Conservation Report No. 1, Egerton University, Njoro, Kenya, 1990.

Paul, S., *Community Participation in Development Projects: the World Bank experience*, Discussion Paper No.6, World Bank, Washington DC, 1987.

Pretty, J.N., *Rapid Catchment Analysis for Extension Agents*, notes on the 1990 Kericho training workshop for the Ministry of Agriculture, Kenya, IIED, London, 1990.

Raintree, J.B., *D and D User's Manual. An Introduction to Agroforestry Diagnosis and Design*, ICRAF, Nairobi, 1987.

Reij, C., *Indigenous Soil and Water Conservation in Africa: an assessment of current knowledge*, Paper to the workshop on Conservation in Africa – Indigenous knowledge and conservation strategies, Harare, 2–7 December 1990.

Rhoades, R., 'Informal Survey Methods for Farming Systems Research', *Human Organization* 44,3, 1985, pp 215–218.

Rhoades, R., and Booth, R., 'Farmer-back-to-farmer: a model for generating acceptable agricultural technology', *Agricultural Administration*, Vol. 11, 1982, pp 127–137.

Richards, P., *Indigenous Agricultural Revolutions: ecology and food production in West Africa*, Hutchinson, London, 1985.

Rocheleau, D., Wachira, K., Malaret, L. and Wanjohi, B.M., 'Local Knowledge For Agroforestry And Native Plants', in *Farmer First, Farmer Innovation and Agricultural Research*, Chambers, R., Pacey, A. and Thrupp, L.A. (eds.), Intermediate Technology Publications, London, 1989.

Rohde, G., 'Kompostierung der Stadtabfalle', *Wissenschaftliche Zeitschrift der Humboldt-Universitat Jhrg. IV*, Berlin, 1954.

Sanchez, P.A., *Properties And Management Of Soils In The Tropics*, chapter 5, Wiley, New York, 1976.

Sanchez, P.A., and Cochrane, T.T., 'Soil Constraints In Relation To Major Farming Systems Of Tropical America', in *Priorities For Alleviating Soil-related Constraints To Food Production In The Tropics*, IRRI, Manila, 1980.

Sands, D.M., *The Technology Application Gap: overcoming constraints to small-farm development*, FAO Research and Technology Paper 1, FAO, Rome, Italy, 1986.

SCRP, *Soil Conservation Research Project Progress Report* CFSCDD, MoA, Addis Ababa, 1986.

Scrimshaw, S., and Hurtado, E., *Rapid Assessment Procedures for Nutrition and Primary Health Care: anthropological approaches to improving programme effectiveness*, United Nations University, Tokyo, UNICEF/UN Children's Fund, and UCLA Latin American Centre, Los Angeles, 1987.

Shaxson, T.F., 'Reconciling Social and Technical Needs in Conservation Work on Village Farmlands', in Morgan, R.P.C. (ed.), *Soil Conservation: problems and prospects*, John Wiley & Sons, Chichester, England, 1981.

Shaxson, T.F., 'Conserving Soil by Stealth', in *Conservation Farming on Steep Lands*, Moldenhauer, W.C. and Hudson, N.W. (eds.), SWCS & WASWC, Ankeny, Iowa, 1988.

Shaxson, T.F., *Conservation Farming Through People's Participation*, Final Report: Project TCP/LES/6755 Government of Lesotho/FAO, Maseru, Lesotho, 1989.

Shaxson, T.F., Hudson, N.W., Sanders, D.W., Roose, E. and Moldenhauer, W.C., *Land Husbandry: a framework for soil and water conservation*, WASWC & SWCS, Ankeny, Iowa, 1989.

Sylwander, L., and Egnell, G., *Soil Conservation for All: an anthropological and technical minor field study of the soil conservation and agroforestry programme in Eastern Province, Zambia*, Working Paper No. 16, Development Studies Unit, Swedish University of Agricultural Sciences, Uppsala, 1989.

Trouse, A.C., 'Soil Physical Characteristics And Root Growth', in *Soil Physical Properties And Root Growth In The Tropics*, Lal, R. and Greenland, D.J. (eds.), Wiley, Chichester, 1979.

Vincent, V., and Thomas, R.G., *An Agricultural Survey of Southern Rhodesia, Part 1*. Government Printer, Harare, Zimbabwe, 1960.

Wortmann, C., 'Effects of Bean Varietal Characteristics on Weed Suppression', *Uganda National Bean Programme, 1990 Annual Report*, 1991.

Young, A., *Agroforestry for soil conservation*, CAB International and ICRAF, Nairobi, 1989.

Further Reading

On participatory rural appraisal (PRA)

Chambers, R., Parry, A., and Thrupp, L.A., *Farmer First: farmer innovation and agricultural research*, Intermediate Technology Pubs, 1989.
RRA Notes 1 to 14 and continuing. Available at no charge on request from: Sustainable Agriculture Programme, International Institute for Environment and Development, 3 Endsleigh Street, London WC1H 0DD, UK.
The *PALM/PRA* series. Available at no charge on request from: MYRADA, 2 Service Road, Domlur Layout, Bangalore 560 071, India.

On land husbandry

FAO, *Soil and Water Conservation in Semi-arid Areas*, Soils Bulletin 57, 1987.
Hudson, N.W., *Soil Conservation*, Batsford, London, 1981.
Hudson, N.W., *Land Husbandry*, Batsford, London, 1992.
Shaxson, T.F., Hudson, N.W., Sanders, D.W., Roose, E., and Moldenhauer, W.C., *Land Husbandry, A Framework For Soil And Water Conservation*, 1989, Soil and Water Conservation Society, Ankeny, Iowa.

From the ground-up case study series

Available from: World Resources Institute Publications, PO Box 4852, Hampden Station, Baltimore MD 21211, USA. Price US$10 each.

No. 1 – Thomas-Slayter, B., Kabutha, C., and Ford, R., *Traditional Village Institutions in Environmental Management: Erosion Control in Katheka, Kenya*, 1991.
No. 2 – Thompson, J., *Combining Local Knowledge And Expert Assistance In Natural Resource Management: small-scale irrigation in Kenya*, 1991.
No. 3 – Dorm-Adzobu, C., and Ampadu-Agyei, O., and Veit, P.G., *Community Institutions in Resource Management: agroforestry by mobisquads in Ghana*, 1991.
No. 4 – Dorm-Adzobu, C., Ampadu-Agyei, O., and Veit, P.G., *Religious Beliefs and Environmental Protection: the Malshegu Grove in Northern Ghana*, 1991.
No. 5 – Tukahirwa, E.M., and Veit, P.G., *Public Policy and Legislation in Environmental Management: Gerracing in Nyarurembo, Uganda*, 1992, May.
No. 6 – Mascarenhas, O., *Indigenous Knowledge and Local Innovators in Resource Management: irrigation in Msanzi, Tanzania*, 1992, October.
No. 7 – Allieu, E.K., Bah, O.M., Lahai, A.C., Sheriff, D.M.A., and Sillah, A.B.S., *Socio-economic Incentives in Environmental Management: forestry for cash crop production in Sierra Leone*, 1992, December.
No. 8 – Clarfield, G., and Lowe, D., *Utilizing Ecological Diversity in Resource Management: examplar land use in Turkana, Kenya*, 1992, December.

Participants And Authors

Key: + = participant, * = participating author,
       # = non-participating author.

Botswana
+ Mr Adam E. Sehuhula, Land and Water Management Programme, PO Box 201142, Gaborone.
Canada
+ Dr Andrew D.R. Ker, IDRC, PO Box 62084, Nairobi, Kenya.
Ethiopia
Mr B.W. Aregay, Community Forestry and Soil Conservation Development Department, Ministry of Agriculture, Addis Abeba.
Dr P.A. Chadhokar, Community Forestry and Soil Conservation Development Department, Ministry of Agriculture, Addis Abeba.
Mr Yohannes G. Michael, SCRP, PO Box 2597, Addis Abeba
Germany
* Mr Hans Diederichsen, SECAP, PO Box 72, Lushoto, Tanzania.
+ Dr Hartmut H. Krugmann, IDRC, PO Box 62084, Nairobi, Kenya.
Ghana
+ Mr Okyeame Ampadu-Agyei, UNDP/FPC, Environmental Protection Council. Ministry of Industry Science and Technology, PO Box M326, Accra.
# Mr C. Dorm-Adzobu, Environmental Protection Council, Ministry of Science and Technology, PO Box M326, Accra.
Japan
+ Mr Masami Yamamori, c/o JICA, PO Box 50572, Nairobi, Kenya.
Kenya
+ Mr Paul Amina, PO Box 49622, Nairobi.
+ Mr Francis K. Chesumbai, Permanent Presidential Commission on Soil Conservation and Afforestation, PO Box 30510, Nairobi.
+ Dr Anthony O. Esilaba, KARI, NARC-Muguga, PO Box 30148, Nairobi.
* Dr Francis N. Gichuki, University of Nairobi, Department of Agricultural Engineering, PO Box 30197, Nairobi.
+ Dr Benjamin M. Ikombo, KARI, PO Box 30148, Nairobi.
# Ms Charity Kabutha, PO Box 44145, Nairobi.
+ Mr P.K. Karimi, Permanent Presidential Commission on Soil Conservation and Afforestation, PO Box 30510, Nairobi.
+ Peter Karinge, KENGO, PO Box 48197, Nairobi.
+ Mr H.G. Kimaru, Presidential Commission on Soil Conservation and Afforestation, PO Box 30510, Nairobi.
* Mr J.T. Arap Leting, Office of the President, Nairobi.
+ Mr Joseph K. Metto, KARI, PO Box 30148, Nairobi.
* Ms Margaret Ngunjiri, PPDU, MOA, PO Box 72670, Nairobi.
+ Mr T.K. Njagi, Permanent Presidential Commission on Soil Conservation and Afforestation, PO Box 30510, Nairobi.
* Mr John W. Njoroge, Kenya Institute of Organic Farming, PO Box 34972, Nairobi.
+ Mr Stephen N.J. Njoroge, Permanent Presidential Commission on Soil Conservation and Afforestation, PO Box 30510, Nairobi.
+ Ms Elizabeth Odour-Noah, National Environment Secretariat, PO Box 67839, Nairobi.

* Mr C.R.J. Nyaga, Director of Forestry, Nairobi.
+ Dr Daniel O. Nyamai, KEFRI, PO Box 20412, Nairobi.
+ Ms Joy Obando, Kenyatta University, Geography Department, PO Box 43844, Nairobi.
+ Mr Mbegera M. Ogwoka, Soil and Water Conservation Branch, Ministry of Agriculture, PO Box 30028, Nairobi.
+ Mr P. Omondi, MAD, PO Box 386, Sare.
* Dr Asenath Omwega, Kenyatta University, Geography Department, PO Box 43844, Nairobi.
+ Mr Peter O. Onyambo, MAD, PO Box 719, Kisii.
+ Ms I. Onyancha, KARI, PO Box 30148, Nairobi.
* Sister D. Rauch, MAD, PO Box 719, Kisii.
* Professor D.B. Thomas, PO Box 14893, Nairobi.

Lesotho
Mr Gedion Shone, SADCC, PO Box 24, Maseru 100.
+ Mr Michael S. Mabaso, Agricultural Department, PO Box 58, Mohales Hoek 800.
# Mr L.C. Weaver, US-AID, Lesotho.

Madagascar
+ Mr Jean Louis Rakatomanana, FOFIFA-DREP, BP 904, Antananarivo.

Malawi
* Mr Nyami Jaffu Mulenga, Ministry of Agriculture, PO Box 30134, Lilongwe.

Republique Rwandaise
* Dr Justin Nsengimana, Ministere d'Agriculture de l'Elevage et des Forets, BP 621, Kigali.

Sudan
* Dr Yagoub Abdalla Mohamed, Institute of Environmental Studies, University of Khartoum, Khartoum.

Swaziland
* Ms Funekile G. Mdluli, Ministry of Agriculture and Co-operatives, PO Box 162, Mbabane.
+ Mr Magalea Ngwenya, Ministry of Agriculture and Co-operatives, PO Box 162, Mbabane.
+ Mr J.D. Vilakati, PO Box 162, Mbabane.

Sweden
* Mr G. Bergman, SIDA/RSCU, PO Box 30600, Nairobi, Kenya.
* Dr Eva Tobisson, Development Studies Unit, Annex 1, Department of Social Anthropology, Stockholm University, S-106 91 Stockholm.

Switzerland
* Dr Hanspeter Liniger, Laikipia Research Programme, PO Box 144, Nanyuki, Kenya.

Tanzania
* Fr Herbert Gappa, Catholic Church, PO Box 26, Bariadi.
* Mr F.B.S. Kaihura, Agricultural Research Institute Mlingano, National Soil Service, PO Box 5088, Tanga.
+ Mr K. Kalaga, Forestry Training Institute Olmotony, PO Box 943, Arusha.
+ Mr A.H. Kishimba, PO Box 9071, Dar es Salaam.
+ Mr Stephen D. Lyimo, Selian Agricultural Research Institute, PO Box 6024, Arusha.
+ Dr Adolpho Mascarenhas, University of Dar es Salaam, Institute of Resource Assessment, PO Box 35102, Dar es Salaam.
* Ms Ophelia P. Mascarenhas, University Library, PO Box 35092, Dar es Salaam.
+ Mr L. Mawenya, Regional Land Use Planning and Soil Conservation Office, PO Box 3163, Arusha.

+ Dr S. Misana, University of Dar es Salaam, Department of Geography, PO Box 35051, Dar es Salaam.
* Dr B. Moshi, Principal Secretary, Ministry of Agriculture, Livestock Development and Co-operatives, Dar es Salaam.
* Mr J.G. Mowo, National Soil Service, Dar es Salaam.
+ Dr Adolf S. Nyaki, Selian Agricultural Research Institute, PO Box 6024, Arusha.
# Dr W. Rugumamu, University of Dar es Salaam, Faculty of Social Sciences, PO Box 35049, Dar es Salaam.
* Mr C.S. Rwejuna, ARDHI Institute, PO Box 35176, Dar es Salaam.
+ Mr Hiza B. Sessiwa, TESO, PO Box 1309, Dar es Salaam.
*Mr Cassian W. Sianga, SECAP, PO Box 72, Lushoto.
+ Mr Alfred C.J. Temba, PO Box 3052, Moshi.

Uganda
* Ms Ruth Kiwanuka, JEEP, PO Box 4264, Kampala.
* Mr Erie S. Tamale, Ministry of Environmental Protection, PO Box 9629, Kampala.
* Dr Eldad Tukahirwa, Institute of Environment and Natural Resources, Faculty of Science, Makerere University, PO Box 7062, Kampala.
* Mrs Joy M. Tukahirwa, Department of Geography, Makerere University, PO Box 7062, Kampala.

UK
* Mr Steven Caiger, High Value Horticulture, Colne House, Highbridge Estate, Oxford Road, Uxbridge UB8 1UL.
* Dr Robert Chambers, Institute of Development Studies, University of Sussex, Brighton BN1 9RE.
* Dr Malcolm G. Douglas, Langenfeld, Easingwold Road, Huby, York YO6 1HN.
* Mr David C.S.J. Hughes, CARE International in Kenya, PO Box 43864, Nairobi, Kenya.
* Dr E. Jones, Department of Research and Specialist Services, Chiredzi Research Station, PO Box 97, Chiredzi, Zimbabwe.
* Mr T. Francis Shaxson, 36 Greenhayes, Broadstone, Dorset BH18 8NA.
* Dr Alistair Sutherland, Mount Makulu Research Station, PB 7, Chilanga, Zambia.
+ Mr Michael Watson, ODA/BDDEA, British High Commission, PO Box 30465, Nairobi, Kenya.
* Dr Adrian P. Wood, Department of Geography, Polytechnic of Huddersfield, Queensgate, Huddersfield HD1 3DH.

USA
+ Ms Anne Ching, Ford Foundation, Rural Poverty and Natural Resources, PO Box 41081, Nairobi, Kenya.
* Mr Gary Cohen, US-AID Africa Bureau, Office of Technical Resources, Room 2669, US Department of State, Washington DC 20523.
# Dr Richard Ford, Program for International Development, Clark University, Worcester, Massachusetts.
* Dr John Lynam, Rockefeller Foundation, PO Box 47543, Nairobi, Kenya.
* Mr Richard Pellek, US-AID, PO Box 30261, Nairobi, Kenya.
# Dr Barbara Thomas-Slayter, Program for International Development, Clark University, Worcester, Massachusetts.
* Dr Diana de Treville, Winrock International, Rt 3, PO Box 376, Morrilton, AR 72110-9537.
* Dr Peter Veit, World Resources Institute, 1709 New York Avenue, Suite 700, NW Washington DC 20006.

Zambia
+ Dr Amos M. Bunyolo, Mount Makulu Research Station, PB 7, Chilanga.
+ Mr Lazarous Chileshe, Mount Makulu Research Station, PB 7, Chilanga.
# Mr Yemba Kaonga, Family Farms Ltd, PO Box 42, Magoye.
* Ms Elizabeth Malayisha, Family Farms Ltd, PO Box 42, Magoye.
* Mr Nawa Mukanda, Mount Makulu Research Station, PB 7, Chilanga.
+ Mr Robby Mwiinga, Natural Resource Department, PO Box 50042, Lusaka.
* Mr Lingston Singhogo, ARPT: Mount Makulu Research Station, PB 7, Chilanga.
+ Mr Blackson Sinyangwe, PO Box 410049, Mungwi, Kasama.

Zimbabwe
Mr T. Busangavanye, Chiredzi Research Station, PO Box 97, Chiredzi.
*Mr S. Kavalo, Department of Natural Resources, PO Box 8070, Causeway, Harare.
* Godfrey Nehanda, Agritex, PO Box 8117, Causeway, Harare.
# Mr E. Mazhangara, Chiredzi Research Station, PO Box 97, Chiredzi.
# Mr P. Nyamudeza, Chiredzi Research Station, PO Box 97, Chiredzi.
* Norman Reynolds, Southern Africa Foundation for Economic Research, PO Box A665, Avondale, Harare.

CGIAR and FAO
* Dr Marcelino Avila, ICRAF, PO Box 30677, Nairobi, Kenya.
* Peter Ewell, The International Potato Centre, Tropical Africa Region, PO Box 25171, Nairobi, Kenya.
* Mr Rodney N. Gallagher, FAO, AGLS, Via Terme di Caracalla, 00100 Roma, Italy.
* Dr J.K. Ransom, CIMMYT, PO Box 25171, Nairobi, Kenya.
# Mr D.W. Sanders, FAO, AGLS, Via Terme di Caracatta, 00100, Rome, Italy.
*Dr Charles Wortmann, CIAT, PO Box 6247, Kampala, Uganda.

***SADCC**
* Dr John Hunter, SADCC-ELMS, Ministry of Agriculture and Marketing, PO Box 24, Maseru 100, Lesotho.
* Mr K.H. Mikael Segerros, SADCC-ELMS, Ministry of Agriculture and Marketing, PO Box 24, Maseru 100, Lesotho.

World Association of Soil and Water Conservation
* Mr R.J. Cheatle, WASWC, PO Box 39042, Parklands, Nairobi, Kenya.
# Professor N.W. Hudson, Silsoe Associates, The Clock House, 2 Bedford St, Ampthill, Bedford MK45 2NB, UK.
+ Dr Maurice Cook, North Carolina State University, PO Box 7619, Raleigh, NC 27695, USA.

Abbreviations Used in the Text

AID	— Agency for International Development (USA)
ARPT	— Adaptive Research Planning Teams (Zambia)
ASAL	— Arid and Semi-arid Lands (Kenya)
CARE	— Co-operative for American Relief Everywhere
CGIAR	— Consultative Group on International Agricultural Research
CIAT	— International Centre for Tropical Agriculture
CIMMYT	— International Maize and Wheat Improvement Centre
CIP	— International Potato Centre
DFA	— Development Fund for Africa
ELMS	— Environment and Land Management Sector (of SADCC)
FAO	— Food and Agriculture Organization (of UN)
FFW	— Food For Work
FISC	— Farm Improvement with Soil Conservation (Lesotho)
FSR	— Farming Systems Research
FSRD	— Farming Systems Research and Development
FSRE	— Farming Systems Research and Extension
GRZ	— Government of the Republic of Zambia
GTZ	— Germany Agency for Technical Co-operation
HASHI	— *Hifadhi Ardhi Shinyanga* (The Greening of Shinyanga, Tanzania)
HVH	— High Value Horticulture
IAR	— International Agricultural Research
IBSRAM	— International Board for Soil Research and Management
ICA	— Intensive Conservation Area (Zimbabwe, Zambia)
ICRAF	— International Centre for Research in Agroforestry
ICRISAT	— International Crops Research Institute for the Semi-arid Tropics
IDRC	— International Development Research Centre (Canada)
IIED	— International Institute for Environment and Development
IUCN	— International Union for Conservation of Nature and Natural Resources (World Conservation Union)
JEEP	— Joint Energy and Environment Projects (Uganda)
JICA	— Japanese International Co-operation Agency
KARI	— Kenya Agricultural Research Institute
KEFRI	— Kenya Forestry Research Institute
KENGO	— Kenya Non-government Organizations
KIOF	— Kenya Institute of Organic Farming
LWM	— Land and Water Management
MAD	— Mobilizing Against Desertification (Kenya)
NARS	— National Agricultural Research Service
NES	— National Environment Secretariat (Kenya)
NGO	— Non-government Organization
NORAD	— Norwegian Office for Regional Aid and Development
NRB	— Natural Resources Board (Zimbabwe, Zambia)
ODA	— Overseas Development Administration (Great Britain)
PPCSCA	— Permanent Presidential Commission on Soil Conservation and Afforestation (Kenya)
PRA	— Participatory Rural Appraisal
PTD	— Participatory Technology Development
R & D	— Research and Development
RMA	— Range Management Areas Project (Lesotho)

RRA	— Rapid Rural Appraisal
SADCC	— Southern-Africa Development Co-ordination Conference
SECAP	— Soil Erosion Control and Agroforestry Project (Tanzania)
SIDA	— Swedish International Development Authority
SWM	— Soil and Water Management
TOT	— Transfer of Technology
UNDP	— United Nations Development Programme
UNEP	— United Nations Environment Programme
USAID	— United States Agency for International Development
WASWC	— World Association of Soil and Water Conservation
WFP	— World Food Programme
WRI	— World Resources Institute

Index of Place Names

Anjouan island, Comoros 49
Ethiopia
  country report 23
  erosion in highlands 23
  farmer attitudes 114
  farmer profiles 115
  Peasant Associations 114, 116
  traditional farming systems 115
  Wolayta district 113
Ghana
  Goviefa – Agodome 165
  population pressure 166
  traditional institutions 167
  Volta region 165
Kenya
  Baringo 113
  Central Highlands 131
  conservation measures 117
  farmer profiles 115, 134–138
  Institute of Organic Farming 130
  Katheka sublocation 155
  Kiambu district 135, 136
  Kisii 191
  Machakos district 131, 155
  Mount Kenya 208
  Muranga district 134
  mwethya groups 156
  Nyanza district 190
  Nyeri district 137
  Siaya and South Nyanza project 49
Lesotho
  Board of Farmers 187
  FISC project 183
  grazing associations 208
  Thabana Morena project 126
Malawi
  country report 26
  research and extension 28

Sudan
  Darfur region 169
  Water harvesting 169
Swaziland
  earth dams 177
Tanzania
  Bariadi 159
  country report 29
  East Usambaras project 47
  extension service 113
  HASHI 113, 161
  Rukwa region 174
  SECAP 113
  Shinyanga district 159
  soil and water management 174
  Sumba-wanga district 174
  water management 174
Uganda
  country report 34
  farmer attitudes 114
  fuelwood planting 172
  JEEP 172
  Kabale district 162
  Mount Elgon project 47
  Nyarurembo village 162
  soil conservation legislation 114, 162
Zaire
  Kibale/Semlike Project 47
Zambia
  agroforestry 203
  country report 37
  Family Farms 203
  Mazabuka district 203
  soil and water management 64, 104
Zimbabwe
  country report 41
  Democratic Company Model 79
  rural structural adjustment programme 79
  vertisol management 197

Subject Index

Acacia albida 204, 205
Acacia senegal 170
afforestation, Kenya 193
   Tanzania 160
   Zambia 39
agroforestry
   by CARE 48
   in Ghana 165
   in Kenya 116
   and PRA 101
   practices 13
   in Zambia 39, 203
aid agencies approach 45
aid donors 54
anthropologists, role of 61

'barefoot vets', Lesotho 188
Board of Farmers, Lesotho 187

CARE 48
cattleposts, Lesotho 213
CGIAR 50
CIAT 51
CIMMYT 52
CIP 53
cluster farmers, Kenya 195
coffee prices 134
community institutions, Ghana 165
community participation 13, 99, 153, 178
composting 133, 136, 139
conservation farming systems 9
conservation measures
   agronomic 122
   agroforestry 124
   choice of 117
   costs and benefits 120
conservation policy
   Zambia 39
   Zimbabwe 44
conservation structures, limits of 6
control structures 6
Country reports
   Ethiopia 23
   Malawi 26
   Tanzania 29
   Uganda 34
   Zambia 37
   Zimbabwe 41
cultural tradition in Kenya 191
deforestation
   Kenya 209
   Malawi 27
   Tanzania 31

   Uganda 172
Democratic Company Model, Zimbabwe 79
desertification, Kenya 192
double digging 133, 136, 138
drama in extension 173, 206

earth dams 155, 177
economics
   and development 71, 251
   and natural resources management 77
   of smallholder farming 250
   of tied ridging 202
economists, role of 250
erosion control, Kenya 156
erosion problem in Ethiopia 23
   in Uganda 114
Eucalyptus, Zambia 207
exclusion of cattle, Ethiopia 25
   Lesotho 212
extension services
   in Malawi 28
   in Tanzania 113
   in Zambia 39, 205

Family Farms, Zambia 203
family planning 115
FAO approach 45
farmer attitudes 114
farmer profiles 115, 134
farmer trials 13, 200
farming systems development 8
farming systems research 33
FISC project, Lesotho 183
flexibility in project planning 113
food supply and demand 16
fuelwood planting
   Kenya 114
   Uganda 172
furrow planting 199

gender in development 4, 62, 79, 113
Gliricidia 50
Grazing Associations, Lesotho 212
grazing rights 188, 212
Grevillea 137
Guatemala grass 50

harambee, Kenya 113
HASHI, Tanzania 113, 161
high-income horticulture 141
history of development 77
   of soil conservation, Kenya 155
   Uganda 114

Zambia 37
Zimbabwe 41
horticulture
  essential oil crops 142
  oleoresins 148
  spices 147
human population, increasing 15
  in Malawi 26

IBSRAM 40, 55
ICRAF 55
IUCN 47
immigration 208
  see also refugees
improving land use 3, 224
institution building 228
  Ghana 165
  Sudan 170
  Tanzania 174
  Uganda 35
  Zambia 109
irrigation committees, Tanzania 174

Joint Energy and Environmental Projects (JEEP) 172

Kenya Institute of Organic Farming (KIOF) 130

labour constraint 231
  in Ethiopia 114
  in Kenya 113
  in Uganda 114
Land Allocation Committees, Lesotho 185
land degradation 4
  in Kenya 113
land husbandry 3, 7
  in Kenya 130
  in Lesotho 126, 128
Leucaena
  in Tanzania 160
  in Zambia 207
livestock
  management, Tanzania 114
  numbers 16, 23
  productivity, Lesotho 217

macro-contour line 114
manpower shortage, see labour
migration
  Ethiopia 23
  Ghana 166
  see also immigration, refugees
mobilizing farmers
  Kenya 191
  Lesotho 186
  Swaziland 180
Model Farm Programme, Kenya 194

mutual help schemes, Ethiopa 115
mwethya groups, Kenya 156

Napier grass 163
national policies 17
new technologies, requirements of 19
NGOs 3, 56

organic recycling 3, 130
overgrazing
  Ethiopia 114
  Lesotho 212
  Malawi 27
  Tanzania 31, 159
ox-drawn ridger 201

Participatory Rural Appraisal (PRA) 83, 87, 91, 226
  for agroforestry 101
participatory soil and water management 99
Participatory Technology Development (PTD) 99, 226
Peasant Associations, Ethiopia 114, 116
pennisetum 50, 163
popular participation 60, 175, 226
population pressure 3, 115
  Ethiopia 23, 115
  Ghana 166
  Kenya 208
  Lesotho 212
  Malawi 26
  Tanzania 30
  Uganda 34, 166
  world 15
poverty linked to degradation 3
programme objectives 17

range management, Lesotho 212
Rapid Rural Appraisal (RRA) 85, 87
refugees
  Ghana 166
  Kenya 208
  Malawi 27
rehabilitation of earth dams, Swaziland 177
research
  CGIAR centres 50–54
  Kenya 208
  Malawi 27
  place of, and record of 16
  Zambia 38
research and development
  Tanzania 32, 113
research and extension
  for agroforestry 103
  Malawi 28
  Zimbabwe 44
runoff control 6

271

rural sociologists, see sociologists
Rural Structure Adjustment Programme, Zimbabwe 79

social issues 57
social mutual help 115
social structures 3, 115
socio-economic
  factors 32, 107
  issues 59
  multidisciplinary approach 104
  perspectives 64
sociologists
  role of 59, 64, 248
soil conservation
  Ethiopia 23
  Lesotho 183
  Tanzania 29
  Uganda 162
soil conservation, history of
  in Kenya 155
  in Zambia 37
  in Zimbabwe 41
soil conservation legislation
  Uganda 162
  Zambia 39
  Zimbabwe 44
soil degradation
  Tanzania 30
  Uganda 34
Soil Erosion Control & Agroforestry Programme (SECAP), Tanzania 113
soil and water management
  Baringo, Kenya 113
  Kenya 117, 190, 208
  Malawi 27
  Tanzania 32, 113
  Uganda 34, 162
  Zambia 37, 39
  Zimbabwe 42, 197
soil productivity
  loss of 6
  improving 19
spring development, Lesotho 189
sannhemp 159
sustainable
  development 153
  farming, Ethiopia 115
  rangeland management, Lesotho 213
  smallholder farming 71, 96, 100, 250
  soil and water management, Malawi 28
sweet potatoes 53

T and V extension, Zambia 205
tea 137
tef 24
tenurial rights 61, 80, 108, 114
  Ethiopia 114
  grazing rights, Lesotho 212
  Kenya, Baringo 113
  Lesotho 184
  Uganda 164
  Zambia 203
terraces
  band 163
  bench 125, 155, 163
  fanya juu 125, 138
  strip 163
theatre in extension 173, 206
theology and land husbandry 160, 193
tied ridges 198
traditional culture, Kenya 191
traditional farming systems
  Ethiopia 24, 114
  Tanzania 176
traditional institutions, Ghana 167
  irrigation, Tanzania 174
training
  needs 231, 235
  programmes 235
tree nurseries
  Ghana 168
  Kenya 192
trus cultivation, Sudan 170

vertisols, Zimbabwe 197
vetiver grass 49, 144
Village Chiefs, Lesotho 183, 185
Village Development Committees, Lesotho 185
village
  institutions 153, 156, 170, 177
  veterinary workers 188
water budget, Kenya 209
water conservation vs soil conservation 20
water harvesting
  Baringo, Kenya 113
  Kenya 133, 138
  Sudan 169
water management, Tanzania 174
water spreading, Sudan 170
women farmers, additional constraints for 11
  as leaders 113, 158
  as beneficiaries 4, 113, 178

www.ingramcontent.com/pod-product-compliance
Ingram Content Group UK Ltd.
Pitfield, Milton Keynes, MK11 3LW, UK
UKHW021833140426
5217IPUK00021B/1422